# Pomegranate Production and Marketing

# Pomegranate Production and Marketing

**İbrahim Kahramanoğlu**
Alnar Pomegranates Ltd.
Güzelyurt, Cyprus
and
European University of Lefke
Faculty of Agricultural Sciences and Technologies
Department of Horticultural Production and Marketing
Güzelyurt, Cyprus

**Serhat Usanmaz**
European University of Lefke
Faculty of Agricultural Sciences and Technologies
Department of Horticultural Production and Marketing
Güzelyurt, Cyprus

CRC Press is an imprint of the
Taylor & Francis Group, an **informa** business
A SCIENCE PUBLISHERS BOOK

CRC Press
Taylor & Francis Group
6000 Broken Sound Parkway NW, Suite 300
Boca Raton, FL 33487-2742

First issued in paperback 2020



ISBN-13: 978-1-4987-6850-4 (hbk)
ISBN-13: 978-0-367-78301-3 (pbk)



**Library of Congress Cataloging-in-Publication Data**

Names: Kahramanoğlu, İbrahim, author. | Usanmaz, Serhat, author.
Title: Pomegranate production and marketing / İbrahim Kahramanoğlu and Serhat Usanmaz.
Description: Boca Raton, FL : CRC Press, Taylor & Francis Group, 2016. | Includes bibliographical references and index.
Identifiers: LCCN 2016004913 | ISBN 9781498768504 (hardback : alk. paper)
Subjects: LCSH: Pomegranate. | Pomegranate--Postharvest technology. | Pomegranate--Health aspects. | Pomegranate industry.
Classification: LCC SB379.P6 K34 2016 | DDC 634/.14--dc23
LC record available at http://lccn.loc.gov/2016004913

**Visit the Taylor & Francis Web site at**
**http://www.taylorandfrancis.com**

**and the CRC Press Web site at**
**http://www.crcpress.com**

# PREFACE

Pomegranate is a traditional crop that has been cultivated for many years. However, because of the inconvenience of extracting the arils (for eating), it had not been very popular. Since 2000, many scientific researches conducted about the health benefits of pomegranates have verified the traditionally known benefits. According to the scientific studies, pomegranate fruit, peel, leaves, flowers and roots contain bioactive phytochemicals that are antimicrobial, reduce blood pressure and act against serious diseases such as diabetes and cancer. The findings of these studies have increased the public awareness of pomegranate and consumption of pomegranate fruit has increased substantially.

Today, pomegranate plant is produced worldwide in sub-tropical and tropical areas and pomegranate fruits are marketed in all over the world. The main problems of pomegranate sector are the difficulties in production and storage during marketing. The book presents up-to-date scientific and theoretical information about production, storage and processing, marketing and therapeutic applications of pomegranates. It presents simple and practical processing and storage techniques for the extending of storage and shelf life of freshly squeezed natural pomegranate juice and retaining its nutritional quality. With the increasing demand for pomegranates and increasing emphasis on healthy eating, this book is crucial to ensuring sustainability for pomegranate production and food safety.

We have practical experience in the production, post-harvest techniques and marketing as well as in promoting consumption of this highly valuable fruit. Much research has been done and many obstacles have been overcome during our studies. We felt that such kind of book is not available; thus we want to share our knowledge with the readers. We hope that you will find the up-to-date, scientific and practical information about pomegranates useful.

# CONTENTS

# LIST OF FIGURES

# LIST OF TABLES

# AUTHORS BIOGRAPHY

**İbrahim Kahramanoğlu** was born in 1984 in Cyprus. He is an expert agronomist and candidating his Doctoral Degree (2016) from the European University of Lefke in Cyprus. Moreover, he has been the managing director of Alnar Pomegranates Ltd. for eight years. He started from the beginning and grew up with the company. He conducted numerous studies about production, post-harvest techniques and marketing of pomegranates. He has several published studies where most of his works are about pomegranates. He is the leading personnel of the company where they produce, pack and export fresh pomegranates and also produce 100% natural, freshly squeezed pomegranate juice.

Tel: +90 533 847 14 71
E-mail: ibrahimcy84@yahoo.com

**Serhat Usanmaz** was born in 1980 in Cyprus and is candidating his Doctoral Degree (2016) at the European University of Lefke. He is a lecturer at the same university and thus responsible for the Research Farm. He has experiments in consulting farmers including pomegranate producers about production and marketing. He is also consulting farmers about GLOBAL G.A.P. certification, food safety, sustainability and organic farming. He has also published some studies about pomegranate production, organic farming and etc.

Tel: +90 542 880 63 73
E-mail: serhat_usanmaz@yahoo.com

# CHAPTER 1

# INTRODUCTION

Pomegranate fruits are categorized within the group of berries. There are many small arils within the fruits wrapped to the inside of leathery peel. Previously, pomegranate plants were thought to belong to their own botanical family, the Punicaceae. Recent molecular studies suggest that the Punica genus might be considered within the Lythraceae family (Graham et al. 2005). There is only one genus within this family, the *Punica*. The *Punica* genus includes two species, with the names of: *P. granatum* L. and *P. protopunica* Balf.

*P. protopunica* is noted to be endemic to the Socotra Island of the Arabina Peninsula and is the only analogous relative of *P. granatum* (Shilikina 1973). Pomegranate was initially named *Malum punicum*, which means the apple of Carthage. Thus, Linneaus called it *Punica granatum*, where granatum means seedy. That's why people in the United States call this fruit as seedy apple. The name *Punica* comes from Carthage, a feminized Roman name, and is an ancient city in Tunisia.

Pomegranates are originated from central Asia (Morton 1987, Holland et al. 2009). But pomegranate trees are adaptable to a wide range of soil and climates. However, it can be grown in many different geographical places including the Mediterranean basin, California and Asia. Pomegranate has an important place in the ancient cultures of the Mediterranean countries. Some scientists declared that the pomegranate is the "apple" of the biblical Garden of Eden, but this idea is rejected in recent studies (McDonald 2002). Edible pomegranates were firstly reported to be cultivated in Iran during 3000 BCE. Phoenicians established Mediterranean Sea colonies in North Africa and brought pomegranates to Tunisia and Egypt By 2000 BCE. During that time, pomegranates also naturalized in Cyprus, Turkey and Greece. Dispersion of the trees continued around the world and reached China by 100 BCE (Stover and Mercure 2007). According to Mortan (1987), cultivation of the pomegranates in the Roman Empire and Spain is estimated to be in 800s and reached Indonesia in 1400s. Reaching of the pomegranates to Central America, Mexico and South America took place in 1500s by Spanish. Pomegranate is among the first five cultivated crops together with figs, dates, olives and grapes. Domestication of pomegranate is reported to have begun in 3000-4000 BC in the North of

Iran and the Himalayas in the Northern India (Lye 2008). According to Langley (2000), "pomegranate represents life, regeneration and marriage in the Greek mythology. In the Persian Wars Herodotus mentions "golden pomegranates" adorning the spears of warriors in the Persian phalanx; in Judaism pomegranate seeds are said to number 613 – one for each of the Bible's 613 commandments; in Buddhism it is one of the blessed fruits and represents the essence of favourable influences; in China it is widely represented in ceramic art symbolising fertility, abundance, and a blessed future; in Christianity it is a symbol of resurrection and eternal life; and in Islam the heavenly paradise of the Quran describes four gardens with shade, springs, and fruits including the pomegranate".

According to Eber's papyrus (1550 BCE), the ancient Egyptians used the root extracts of pomegranate for the riddance of tapeworms (Wren, 1988). On the other hand, Hippocrates (400 BCE) also used extracts of pomegranate for many purposes, i.e. eye inflammation and aid to digestion (Adams 1849). In Mexico, extracts of the pomegranate flowers were used to relieve mouth and throat inflammation (Morton 1987). Recent scientific studies confirm traditional usage of the pomegranate fruit and tree for medical purposes. Pomegranate fruit, flowers, bark and leaves contain beneficial phytochemicals which are anti-oxidant and anti-microbial, reduce blood pressure and act against serious diseases such as cancer and diabetes (Jurenka 2008). Lots of studies carried about pomegranates (Gil et al. 2000, Aviram and Dornfeld 2001, Lansky et al. 2005, Türk et al. 2008, Haidari et al. 2009) and confirmed the benefits of pomegranate for human health. Valuable results of these studies increased the public awareness about pomegranate and thus consumption of pomegranate fruit increased throughout the world. However, consumption is still limited due to the hassle for extracting arils. Thus, damages by many pests and diseases and occurrence of physiological disorders such as cracking and chilling injury are other challenge which affects production and reduces marketability of the produce. Moreover, many physiological, biochemical and textural damages occur post-harvest resulting changes in color, taste, texture and reduce nutritional quality too.

Nowadays, pomegranate plant is produced throughout the world in sub-tropical and tropical areas. Mediterranean countries, India, Iran and Californian are the main producers. Argentina, Israel, Brazil, Peru, Chile and South Africa are the other important producer countries. The genetic diversity of pomegranate is demonstrated by an excess of 500 globally distributed varieties, approximately 50 of which are known to be commercially cultivated (IPGRI, 2001). India, China, Iran and Turkey have the largest area of production. Iran and India are the greatest exporters. No clear data is available about the total area and total production in the world due to the rapid increase in the production. Total production in 2014 is estimated to be around three million tons.

## The Plant

Pomegranate trees are known to be long-lived plants, however the fruit production declines after fifteen years. Trees are shrub-like and can grow up to 9 meters in height. However, they generally grow up to 3-4 meters especially in cultivated areas (Fig. 1). Some dwarf varieties are also available. The pomegranate plant is more or less spiny. Bark of the tree is brown-reddish in color when the plant is young and turns to grayish tone as it matures. There is a high tendency to sucker from the plant base. In orchards, plants are normally trained to a single trunk. Trees may be trained to multiple trunks too. Scientists are divided into two groups, one group is suggesting one trunk and the other group is suggesting multiple trunks. The important point here is the quality of the fruits. Since the tree is spiny, multiple trunks cause more damages on the fruits where the trees become crowded. Another important point is the distance among trees and number of trees for unit area. Actually, if the number of trees per unit area is well-organized; and number of fruits per tree are thinned to suitable number according to tree power, no significant difference is found between the yields of neither one trunk nor multiple trunks. Most pomegranates begin fruiting in their third year but productive yielding begins after 5 yr. The yield of a tree may be about 10-20 kg in third year and may reach up to 60-100 kg in fifth year, depending on the cultivar, geographical region and production practices as well.

**Figure 1.** A view of a pomegranate tree

## The Flower

Pomegranate flowers occur about one month after bud appearance on newly developed branches of the same year. Flowers generally occur on spurs. This information is very important especially for pruning practices. Wrong pruning practices can result in no flower. Flowers can appear solitary, in pairs or in clusters. Flower calyx is tubular, with 5 to 8 lobs and fleshy. Flower petals are 5-7, brilliant-red in color and lanceolate. Flower stamens are numerous, free, borne on calyx tube and filaments free (Fig. 2). The ovary is inferior with several locules (Holland et al. 2009).

**Figure 2.** A view of a pomegranate flower

Pomegranate flowers are usually red, orange or pink in color. White flowers have also been described (Newman and Lansky 2012). They are more than 3 cm in diameter. First sign of the fruit is the crown-like protuberance at the flowers base. Flowers are generally 'perfect', containing both male and female parts. However, the same pomegranate plant can carry three types of flowers; namely hermaphrodite, male and intermediate forms. Although the flowers are generally self-pollinate, insects may require for thorough pollination. Some pomegranates are not self-fertile and require compatible pollinizers. Pollination may be performed by insects, usually honey bees. Flowering in northern hemisphere generally occurs from February to April and in southern hemisphere from July to August. It continues for up to 3 months or more depending on cultivar and geographical condition.

The period of full bloom lasts about one month, and fruit set occurs in about 2 or 4 distinct periods. But, the high quality fruits obtained from the first distinct which took place 3 to 4 weeks after the onset of blooming (Fig. 3). Thus, thinning the late fruit can improve the fruit quality. The fruit of first-flower is generally the fruit which appear on the tip of the branch. After overall fruit set (when fruits reach about walnut size), thinning of the smaller and axillar fruits may be performed to increase fruit quality. This also helps to control some pests, like Mealy bugs.

**Figure 3.** A view of fruits in clusters

## The Fruit

Pomegranate fruits are flattened from the up and down or globose, 5-18 cm in diameter and has a leathery skin. The inside of fruit is divided by thin inedible membranes into a number of cells. These cells are packed full of arils. The arils are juicy and include seeds which are although considered inedible but consumed by human beings nevertheless (Teixeira da Silva et al. 2013). Seeds are high in fiber and oil and are beneficial for human health. Fruit ripens about 4-7 months after flowering. High temperatures during the fruiting period are beneficial to have best flavor. Generally, darkly color fruits have the best flavor. Increase in the temperature differences between day and night, improves the color and flavor of the fruits. The weight can vary from 200 grams to 1000 grams. However, few fruits can grow up to be 1800 grams (Fig. 4). The outer peel of the fruit is bitter. This is also rich in compounds with potential therapeutic uses. These include tannins and flavonoids which are natural preservatives of powerful antioxidants (Newman and Lansky 2012).

**Figure 4.** A view of a fruit with 1780 g

## Arils/Seeds

The peel of the pomegranate fruit is not edible. Inside of the fruit is filled with arils including juice and seeds. The inside of the fruit is divided into many chambers packed with the arils (Fig. 5). The taste varies from sweet, tart-sweet to sour. "In Judaism pomegranate seeds are said to number

**Figure 5.** A view of a star-cut pomegranate fruit with chambers including arils

613 – one for each of the Bible's 613 commandments" (Langley 2000). But, the real number of arils is generally between 200 to 800 depending on the variety and size. This can be as high as 1300 (Stover and Mercure 2007). When the juice is removed from the arils, the seeds are oil reach which can be used as a skincare aid. About 18% of dry weight of the seeds is oil. About 65% of the oil is punicic acid. It is a triple conjugated 18-carbon fatty acid (Newman and Lansky 2012).

## The Juice

Juice content of the fruits can vary from 20% to 50%. Pomegranate juice, as with all fruit juices, has its share of natural sugar (fructose, glucose and sucrose) and simple organic acids (ascorbic, citric, fumaric and malic). The juice also contains essential amino acids (Newman and Lansky 2012).

# CHAPTER 2

# IMPORTANT CULTIVARS

Hundreds of cultivars exist across many countries throughout the world. Pomegranate cultivars can be classified according to:

- Taste: sweet, sweet-sour, tart and sour.
- Harvesting time: early, mid-season and late.
- Consumption pattern: juice and table fruit
- Seed hardness: soft-seeded and hard-seeded

The names originate frequently either from the place of cultivation or from the color of the fruit. The genetic diversity of pomegranate is demonstrated by an excess of 500 globally distributed cultivars, nearly 50 of which are known to be commercially cultivated (IPGRI, 2001). Sour varieties are not preferred by most of the consumers. Hard-seeded varieties are also less preferred. These varieties are generally used for processing if the juice color and content is good.

## Wonderful

This cultivar originated in Florida. It has large, deep purple-red fruits (Fig. 6). The fruit rind is medium thick. The inside is deep crimson in color, juicy and has a tart taste. Its seeds are not very hard. This cultivar is better

**Figure 6.** A view of a Wonderful cultivar pomegranate fruit

for fresh eating and juicing. Seperation of arils is easy. Plant is vigorous and productive. Leading commercial cultivar is California and is the most popular cultivar in the world. Fruit sizes: 350-600 grams. Ripening Brix: 17-21. It requires about 160-180 days from flowering to ripening.

## Hicaznar

It is the most known Turkish cultivar. Internal and external color is dark; shelf life is good and suitable for storing and transportation. Fruits are medium to good size and medium-late variety (Fig. 7). Taste is tart and seeds are hard. Separation of arils is not easy because of cohesive fruit flesh. Fruit sizes: 300-450 grams. Ripening Brix: 16-18. It requires about 170-190 days from flowering to ripening.

**Figure 7.** A view of a Hicaznar cultivar pomegranate fruit

## Bhagwa

It originates in Maharashtra, India. Its fruits are large with yellowish red peel and pinkish aril with soft seeds (Fig. 8). Fruit sizes: 200-350 grams. Ripening Brix: >16. It requires about 150-180 days from flowering to ripening.

**Figure 8.** A view of a Bhagwa cultivar pomegranate fruit

## Arakta

The cultivar originates in India. It is a soft-seeded cultivar and known to be mid-maturing. The skin color is bright ruby red and arils are rose pink in color. Juice content is thought to be high where seeds are soft. Fruits are medium to large sized (Fig. 9). The variety favors a semi-arid climate and is known to be drough tolerant. Fruit sizes: 225-375 grams. Ripening Brix: 18. It requires about 150-180 days from flowering to ripening.

**Figure 9.** A view of Arakta cultivar pomegranate fruit

## Malas Yazdi

It is the well-known variety of Iran. The pomegranate with a thick reddish color skin and a round shape has a size similar to an orange or sometimes as large as a grapefruit. Skin color and the inside color is pinkish to red in color (Fig. 10). Arils contain white seeds. Taste is sweet and sour. It requires about 160-180 days from flowering to ripening.

**Figure 10.** A view of a Malas Yazdi cultivar pomegranate fruit

## Mollar de Elche

Mollar de Elche is the best-known Spanish cultivar with sweet taste fruit and soft seeds (Fig. 11). The peel color is pink-red where the arils are red in color. Generally ripens in October-November in Spain. Large or very large sized fruit from 300 to 600 grams. It is vigorous and fast growing tree. Ripening brix is about 14-17. It requires about 150-180 days from flowering to ripening.

**Figure 11**. A view of a Mollar de Elche cultivar pomegranate fruit

## Acco

It is a very sweet Israeli cultivar with pinkish skin and pulp (Fig. 12). Seeds are softer than Wonderful and many other standard cultivars. The flowers are attractive pinkish-orange in colour. Very juicy and easy separation of arils. Better for juicing and eating. Fruit sizes: 250-450 grams. Ripening Brix: 17-18. It requires about 130-150 days from flowering to ripening.

**Figure 12.** A view of a Acco cultivar pomegranate fruit

## Herskovitz

This cultivar is originated in Israel with sour taste. Red skin and pulp (Fig. 13). Seeds relatively hard. Attractive pinkish-red flowers. Very juicy and easy separation of arils. Better for juicing and eating. Fruit sizes: 250-450 grams. Ripening Brix: 15-17. It requires about 130-150 days from flowering to ripening.

**Figure 13.** A view of a Herskovitz cultivar pomegranate fruit

## Other Cultivars

Apart from the above listed 8 cultivars, followings are the other important cultivars worldwide (Holland et al. 2009):

- *China:* 87-Qing 7, Heyinruanzi, Tongpi, Dabaitian, Teipitian and Bopi.
- *Egypt:* Arabi, Manfaloty, Nab El Gamal and Wardy.
- *India:* Ganesh, Mridula, Jalore, Alandi, Muskat, Jodhpur Red, Dholka, Malta, Kandhari, Chawla, Nabha and Achikdana.
- *Iran:* Malas-e-Saveh, Rabab-e-Neyriz, Sishe Kape-Ferdos, Naderi-e-Budrood, Ardesstani Mahvalat, Bajestani Gonabad, Ghojagh Ghoni, Khazr Bardaskn, Bajestan, Zagh, Togh Gardan and Esfahani Daneh Ghermez.
- *Morocco:* Gjeigi, Grenade Jaune, Gordo de Javita, Djeibali and Onuk Hmam.
- *Turkey:* Çekirdeksiz, Ernar, Canernar, Fellahyemez, Hatay, İzmir 1, Katrbas and Silifke Narı.

# CHAPTER 3

# ECOLOGICAL NEEDS

## Climate

Pomegranate plant is adapted to climates with cool winters and hot summers. Thus it is accepted as mild-temperate to subtropical fruit. But, the plant can adapt to a wide range of climatic conditions. Plants can be severely injured by temperatures below –11 °C. It might grow from the plains to an elevation of nearly 2000 m. However the economic production is difficult over 600 m. High temperature is beneficial especially during ripening period. This causes fruits to become sweeter and reddish in color depending on the variety. Thus, increase in the temperature differences between day and night also improves the fruit flavor. Humid climate negatively affects fruit quality. Pomegranate tree can withstand drought, but the yield decreases too much. Plant leaves are susceptible to frost.

## Soil and Water

The pomegranate trees have good adaptability to varying soil conditions except saline, very calcareous, alkaline and sandy soils. Pomegranate tolerates mildly alkaline soils (up to pH 7.5) but prefers slightly acid soil (pH 6.0-6.5). Pomegranate trees are known to be resistant to salinity. However, only the tree is resistant. The tree can withstand the salinity but the fruits may be damaged. Considerable yield decrease is seen on saline soil and water conditions. The water salinity should not exceed 3.500 ppm. Trees grow best on deep, rather heavy loam and alluvial soils. Trees can tolerate soils which are limy and slightly alkaline. Light to sandy soils can also be used. However, the yield is not high in this type of soil even if the orchards are well-irrigated.

# CHAPTER 4

# PRODUCTION

## Propagation

Reproduction of pomegranates is generally done by cuttings. Reproduction of new plants from the seeds is not preferred due to the non-uniform fruits of plants developed from seeds. The fruit quality and fruit yield of those trees are also low. Grafting is another reproduction method for pomegranates. However, the trunk of the pomegranate trees is not hard and the bark can easily swell. Those cause the scions to break easily when the trees bear fruits. Therefore, it is not preferred to use grafting for the reproduction of pomegranates. The best propagation method for the reproduction of pomegranates is by cuttings. This is very important to produce true-to-type cultivars. The cuttings, preferably 10 to 15 cm long, should be obtained from one year old, fully mature wood. The best time of making the cutting is the time when the plants shed leaves before opening. The cuttings are planted in the nursery fields and usually become ready for transplanting within 6 months. During this period, cuttings need to be regularly irrigated. Until the appearance of leaves, daily irrigation is needed. Thus, irrigation could be performed with 2 days interval. Root suckers can also be used to reproduce new plants.

## Orchard Establishment

Before planting, it is highly recommended to deeply plough soil. Then it is beneficial to cultivate with harrowing. Planting is done in well-prepared pits of about 30-50 cm deep. The planting of the rooted cuttings to the field may be done at the beginning of the season, before flushing. It is also recommended to prune side shoots and long roots before planting because they may reduce the power of the cuttings. Plants get all their energy from the sun to grow and develop. Thus, orientation of the orchard needs to be assigned properly to ensure better sun penetration. To achieve maximum fruit quality and yield, sunlight must be properly distributed within the orchard. Orientation of the tree rows should be in north to south for best light interception, whenever possible. The potential light interception by trees in north/south rows is always higher than the equivalent trees grown in east/west rows.

Another important factor which is affecting the light penetration is the planting density (distance between plants and rows) and tree height. The preferred distance among the plants is 3 m (or 2 m) and among the rows is 5 m (or 4 m), but it is depending upon the variety of pomegranate. For the dwarf varieties, 1 m among plants and 3 m among rows may also be used. However, this type of production needs trellis system to support the trees. Light irrigation should be given shortly after planting.

## Pruning

Trees may be trained to a bush, multiple or single stem. The bush form may only be used for backyards, and it is undesirable for economic production. Pomegranate trees naturally tend to produce many suckers from the base. If a single stem system is desired, one vigorous stem (or sucker) is selected and branches grown from it. Basal suckers must be removed regularly (Fig. 14). Because these suckers take the energy of the plant and reduce yield and quality. On this main stem, 3-5 main branche(s) are allowed to grow nearly 60-80 cm from the ground level.

**Figure 14.** Removing of suckers from the tree base

If multiple stem system is desired 3-6 vigorous suckers are selected around the base of the young tree and allowed to grow. Selection of the stems may take two or three years until good trunks are selected correctly distributed around the tree base. All other suckers should be removed regularly. Due to the sensitivity of pomegranate trees to frost injury, farmers may choose multiple stem system in climatic conditions where frost injury is possible. This is because in case of frost damage, generally half of the trunks are damaged and rest stays undamaged. Thus, new stems can be trained from suckers in 2-3 years. In such case for single stem system, whole orchard may be lost.

For both systems, tying of the stems (to each other in multiple stem system and to a support in single stem system) is required for support for the first 2-3 years. Next, stems become rigid enough to carry the canopy and the yield. Pomegranate trees require pruning each winter to ensure proper number of spurs for fruit yielding. Branches from the main stem or stems need to be pruned back from 40-50 cm. This is therefore of paramount importance to ensure that the trees carry fruits. Fruits are borne on short branches which are called spurs. The older spurs lose the capacity to bear fruit and need to be removed. It is desirable to encourage new growth on 1 to 3 year old wood. The short spurs on 2 (sometimes 3) year old wood produce flowers. Light, annual pruning encourages growth of new fruit spurs (Fig. 15). However, heavy pruning reduces yields where wrong and heavy pruning can result in no yield. Therefore, pruning should be performed to leave adequate number of fruit-bearing wood on the tree. Removal of crossing over or interfering branches is necessary. Moreover, thinning out of crowded branches helps to produce larger fruits with less or without wind damages.

**Figure 15.** General view of young pomegranate tree for showing single stem pruning system

Pruning must be performed during the winter months after cold weather but before budding. Thus, pruning needs to be continued regularly by removing of suckers from the base and shortening of the branches from 40-50 cm. Pruning should be aimed to ensure vase-shaped tree with enough branches to have yield and support the tree. Another aim should be not to prevent airflow and sunlight penetration into the canopy.

Sometimes, especially during the first 3-4 years, branches may need support to ensure that fruits do not touch the orchard soil. Trees should be maintained at a height not exceeding 3 m to minimize work for harvesting. Excessive pruning may result in more sun damage of the fruits. Trees must have adequate number of branches and leaves to protect fruits from sunburn.

Pomegranate fruit is very sensitive to sunburn. Pruning and shaping of the tree is very important in order to prevent sunburn damages. Pruning must aim to stimulate trees to produce fruits underside of the tree to prevent sunburn damage. Vertical single wire, vertical double wire or V trellis systems may be used to support pomegranate trees in need of support and thus to produce shade

## Irrigation

Regular irrigation for the establishment of young plants is recommended. Recommended irrigation period is between 1 to 7 days depending on the climate and soil condition. On sandy soils, irrigation need to be performed daily, but on clay-loam soils, irrigation interval may be up to 7 days during spring and maximum 3 days during extreme summer conditions. The fruit cracks if there is irregularity in irrigation during fruit development. The general water requirement of pomegranate trees is given in Table 1. These requirements are for the Mediterranean climate and small

**Table 1.** Daily water requirements of pomegranate (l/tree)

| Months | 1 yr | 2 yr | 3 yr | 4 yr | 5+ yr |
|---|---|---|---|---|---|
| **Month 1** | 4 | 7 | 10 | 11 | 13 |
| **Month 2** | 6 | 12 | 17 | 19 | 21 |
| **Month 3** | 8 | 14 | 19 | 22 | 27 |
| **Month 4** | 10 | 17 | 21 | 28 | 40 |
| **Month 5** | 12 | 20 | 26 | 32 | 50 |
| **Month 6** | 12 | 20 | 26 | 32 | 45 |
| **Month 7** | 10 | 18 | 24 | 28 | 40 |
| **Month 8** | 7 | 15 | 20 | 24 | 30 |

*Month 1 is the month of waking up (bud breaking) which correspond to February or March in northern hemisphere and September or October in southern hemisphere

changes are needed for different climates. In general, every 1°C increase in temperature (in comparison to the Mediterranean climate) causes an increase in the irrigation requirement of nearly 2 l more.

## Fertilization

Application of well-rotted farmyard manure is recommended @ of about 5 kg per tree at the time of planting. Fertilizer requirements of the pomegranate trees change depending on the variety and age of the trees. The pure Nitrogen, Phosphorus and Potassium requirements of the 2, 3, 4 and 5+ year pomegranate trees are given in Table 2.

**Table 2.** Pure fertilizer requirements of pomegranates (gr/tree/yr)

| Age of tree | Nitrogen (N) | Phosphorus (N) | Potassium (K) |
|---|---|---|---|
| **2 yr old** | 100 | 50 | 80 |
| **3 yr old** | 150 | 80 | 130 |
| **4 yr old** | 250 | 120 | 200 |
| **5+ yr old** | 300 | 180 | 250 |

Nitrogen is very important for cell enlargement & division and shoots development. Therefore, amount of nitrogen is very important in determining the maximum size of the fruits by increasing the number of cells inside the fruits. This fertilizer must be applied throughout the growing season, but it is recommended to stop applying before harvest. Nitrogen application before harvest may improve fruit cracking. Phosphorus is necessary for the rooting and flowering. The annual requirements must be applied during flowering for about 4 months. Potassium is required for the enlargement of the cells. It is recommended to apply after fruit set until harvesting. Apart from the above mentioned fertilizers, about 50 kg well-rotted farmyard manure and 1 kg ammonium sulphate prior to flowering are recommended for healthy growth and quality fruiting.

Occasionally zinc, calcium, manganese, magnesium deficiency is evident in pomegranate tress. It is highly recommended to apply zinc, calcium, manganese and magnesium sprays during foliage in spring and mid-summer. These micronutrients, especially zinc and calcium are very important for the prevention of fruit cracking. Application of fertilizers must be performed according to leaf and soil analysis. Leaf samples may be collected two months after bud breaking from the non-fruit bearing terminals. The leaves should be about 3 months old from spring cycle. The optimum leaf mineral concentrations are given below:

| | |
|---|---|
| Nitrogen | : 1.4-2.0% |
| Phosphorus | : 0.1-0.2% |
| Potassium | : 0.6-1.0% |
| Calcium | : 0.6-2.4% |
| Magnesium | : 0.3-0.5% |
| Manganese | : 30-70 ppm |
| Zinc | : 20-60 ppm |
| Iron | : 40-100 ppm |
| Boron | : 10-20 ppm |
| Copper | : 10-20 ppm |

# CHAPTER 5

# POMEGRANATE PESTS

## Aphids

Aphids (*Aphis punicae, Aphis pomi*) are among the most important and widespread pests in pomegranate orchards. They generally feed on the young leaves which are highly susceptible to aphid attacks. High humidity favours the multiplication of aphids where extreme temperatures cause them to die. This pest is distributed throughout the Mediterranean region and peak during April – July, depending on the climatic conditions.

### Identification

Aphids are yellowish-green, green, yellow, brown or black depending on the species. But, the *Aphis punicae* is generally yellowish-green in color. They have soft pear-shaped bodies with long legs. Some species of aphids can have a waxy coating over their bodies which make it difficult to control them (UC IPM 2015). Aphids are soft-bodied insects with long mouth parts. They use these mouth parts to feed on small branches, leaves, flowers (Fig. 16) and small fruits by sucking out fluids. Some adult aphids are wingless where most species have wings. This characteristic causes aphid populations to disperse quickly.

**Figure 16.** A view of damage on flowers caused by Aphids

### Life Cycle

Aphids may have many generations throughout a year. Some aphid species produce sexual forms where they mate and produce eggs in fall. In this case, there is no foliage on the trees for eggs and thus aphids lay their eggs on an alternative host; usually on the weeds or nearest evergreen trees. When the weather becomes warm, these eggs develop into nymphs and become adults in about one week from newborn nymph. Each adult aphid may produce 10-20 nymphs per day and they may live for about 3 weeks. This causes aphid populations to increase at a great speed (UC IPM 2015). On the other hand, some aphids may reproduce asexually where adult females give birth without mating. Nymphs, the young aphids, shed their skin about four times before becoming adult. They do not have any pupa stage like many other insects.

### Damages

Aphids spend the winter on weeds, near trees and sometimes on pomegranates. In spring, aphids rapidly develop on young leaves, shoots, flowers and young fruits. If heavy populations last more than 2 weeks on trees, stunted young or weak trees are observed. Since they feed on leaves and on fruits, they produce sticky exudate (honeydew) on leaves and fruit, which serve as substrate for some fungus. This exudate turns black with the growth of a sooty mold fungus (Fig. 17). Aphids may also transmit viruses from other plants or between pomegranate trees.

**Figure 17.** A view of damage on fruits caused by sooty mold after Aphids

### Management

Few populations of aphids do not produce a problem for pomegranate farmers. However, if the populations increase, which is very easy for the aphids, it becomes a problem for farmers. For a general perspective, it is well-known that chemical control with insecticides may destroy beneficial insects too.

### Biological Control

Natural enemies, especially lady beetles, can control the aphids if the population is low. However, in reality it is not very effective in controlling aphids. Aphids and ants are known to have a symbiotic life. During the wingless stage of aphids, ants carry them onto the trees and the aphids, in return, produce honeydew for the ants. Ants also protect the aphids from some natural enemies. This knowledge may also be used for the control of aphids where control of ants cause to delay the infection of aphids. To control ants, sticky materials may be applied around the trunk of older trees which prevents the ants from climbing on trees. However, it is suggested to not apply the sticky materials directly on the tree. The trunk may be covered with a wrap and sticky material may be applied onto the wrap.

### Cultural Control

Removing the source of aphids, like weeds is a very important cultural control for the aphids. Removing of weeds must be performed before trees wake up. Removing of the pomegranate suckers from the base is another important cultural control method. On the other hand, it is important to keep in mind that aphids prefer young leaves. Therefore, high amount of nitrogen application may favor aphids.

### Chemical Control

Pomegranate orchards need to be checked regularly, at least at 3 days intervals during the spring. This is very important for the control of aphids in order to determine infestations at the correct time and apply insecticides if needed. The population of aphids generally increases when the temperature is 20-30°C. If population increases and the leaves begin to curl, it will become difficult to control aphids. Followings are some important insecticides which have been tested and known to be effective against aphids (Table 3).

**Table 3.** Available active ingredients for the control of aphids

| Active ingredient | Suggested dose | Harvest interval | EU MRL (ppm) |
|---|---|---|---|
| Acetamiprid 20% | 40 ml/100 l water | 14 days | 0.01 |
| Thiamethoxam 141g/l + Lambda cyhalothrin 106 g/l | 40 ml/100 l water | 14 days | 0.05 & 0.02 |
| Thiacloprid 240 g/l | 40 ml/100 l water | 14 days | 0.02 |
| Imidacloprid 200 g/l | 20 ml/100 l water | 14 days | 1.00 |

All of the above listed insecticides are very effective in controlling aphids. Sometimes one application is adequate to control pests. However, if the weather condition stays warm (not hot), second application may be necessary for best control. At that time, selection of another active ingredient is very important to prevent resistance in aphids against insecticides. Rouhani et al. (2013) conducted a study on the chemical control of *Aphis punicae* in pomegranates in Iran. He evaluated the mortality of Aphids in laboratory conditions against imidacloprid, thiamethoxam, thiacloprid and flonicamid. They reported that the LC50 value for imidacloprid, thiamethoxam, thiacloprid and flonicamid were found to be: 0.24 µl/ml, 0.31 mg/ml, 0.48 µl/ml and 0.05 mg/ml, respectively. Thus the sensitivity of the insects to the tested pesticides were recorded to be imidacloprid > thiacloprid > flonicamid > thiamethoxam.

## Butterflies

One of the most important butterfly species that damages pomegranates is the Anar butterfly (*Virachola isocrates, Virachola livia* or *Deudorix isocrates*). This species is located within the Lepidoptera order under the family of Lycaenidae. It is also known as pomegranate fruit borer or pomegranate butterfly. It is a widespread and destructive pest distributed all over Asia.

### Identification

This pest has dark brown, short and stout larves. They are 1.5-2 cm long. Adults are bluish brown in color (Fig. 18). Females have "V" shaped patch on their forewing.

### Life Cycle

Adults lay their eggs singly on leaves or flower buds. Incubation period lasts 7-10 days and larval period lasts for 20-50 days. Pupa period may occur inside the host fruit (damaged crop) or on the branch. Pupa period lasts for 5-35 days. Total life cycle is completed in 1 to 2 months (UC IPM 2015).

**Figure 18.** A view of an Anar Butterfly

## Damages

Anar butterflies are among the most important and destructive pests of pomegranate. Adults lay their eggs into the flower buds and the caterpillar bores into young fruits. They feed on the internal contents of the fruits and cause rots inside the fruit. When larva grows, it comes out by boring through the fruit peel (Fig. 19).

**Figure 19.** A view of damage on fruits caused by Anar Butterfly

### Management

No biological control is known against this pest. Cultural control is also difficult but it should be kept in mind that clean fields are always essential for decreasing of the pest damages. Damages by the pest generally occur during flowering period. During that time, orchards need to be controlled regularly and presence of more than 5 butterflies for 1 ha area is a reason for chemical application.

Followings are the tested and successful insecticides against anar butterfly (Table 4). Indoxacarb application is more effective than Spinosad and Bacillus thuringiensis (BT). But, it is highly recommended to use Spinosad and BT within the application programs which are biological. Sometimes, more than one application is needed for the effective control of pest.

**Table 4.** Available active ingredients for the effective control of Anar butterfly

| Active ingredient | Suggested dose | Harvest interval | EU MRL (ppm) |
|---|---|---|---|
| Indoxacarb 150 g/l | 35 ml/100 l water | 14 days | 0.02 |
| Spinosad 480 g/l | 20 ml/100 l water | 3 days | 0.02 |
| *Bacillus thuringiensis* 32000 IU/mg | 100 g/100 l water | N/A | No MRL Required |

Morton (1987) reported that this pest lays eggs on flower-buds and the calyx of developing fruits and they may cause loss of an entire crop unless the flowers are sprayed 2 times 30 days apart. Kahramanoğlu and Usanmaz (2013) reported that *D. livia* is the highest damaging pest of pomegranates in Cyprus with a damage of 14.63% in 2011 and 15.57% damage in 2012 in the untreated control treatments. Ksentini et al. (2011) reported that the damage caused by *D. livia* in Tunisia pomegranate varieties is between 5.2% and 52%. Findings of present study are among these limits. Blumenfeld et al. (2000) reported that *D. livia* lay its eggs into crown and damages occur during ripening or in storage and they reported that *Bacillus thuringiensis* is the best way to control this insect. During recent study, *Bacillus thuringiensis* reduced the damage from 15% to below 5% but the highest effect was obtained from the indoxacarb treatment. Singh and Singh (2000) also reported that *Bacillus thuringiensis* is an effective control agent for the *D. livia*. Kahramanoğlu and Usanmaz (2013) reported that indoxacarb is very effective in controlling *D. livia* with only 1.46% damage in 2012. However, they reported that, at least 2 applications are needed. They also noted that *Bacillus thuringiensis* and spinosad are effective biological control agents for the *D. livia* especially in organic farming systems, where the damage may decrease from 15% to below 5% with the application of these biological pesticides.

## Moths

Carob moth (*Ectomyelois ceratoniae*) and Honeydew moth (*Cryptoblabes gnidiella*) have similar behavior and look similar. Moths can have silver and black forewings and wings may be cinnamon brown and black.

### Identification

Eggs are oval and irregularly reticulate. At the beginning they are white in color and become yellow before hatching. Larvae are about 12 mm long when they grow and narrowing towards the end. Head is reddish-brown to black in color (Fig. 20). Pupae are about 6 mm long, greenish in color at the beginning and turn to reddish-brown. Adults are about 1-2 cm. Head and thorax are greyish-brown in color. They have simple antennae (UC IPM 2015).

**Figure 20.** A view of a Carob Moth

### Life Cycle

Adults are known to be active at night and mate on the same night of emergence. And females begin oviposition the following day. Adults are generally attracted to sweet materials, like honeydew excreted by aphids or mealybugs. Adults live for about 10 days during warm periods and for about 30 days during cold periods. Females lay their eggs on fruit and the eggs hatch after about 5-15 days depending on the climate, fewer days in warmer climates. When larvae hatch, they feed on the honeydew and on maturation, they begin to feed on fruits.

**Damages**

Moths sometimes attack healthy pomegranate fruits. However, increase in the number of pests may cause serious damages. Moths prefer to lay eggs on damaged or cracked fruits. Feeding inside the food may cause a fruit rot similar to that of the damage by Anar butterflies.

**Management**

Prevention of fruit damage is very important for the control of moths. Removing previous season's crops from the trees and elimination is also important to eliminate the hosts for pests. Following insecticides are effective against the moths. One application may be enough to control this pest. Economic threshold is about 10 individuals per 1 ha area.

**Table 5.** Available active ingredients for the effective control of Moths

| Active ingredient | Suggested dose | Harvest interval | EU MRL (ppm) |
|---|---|---|---|
| Indoxacarb 150 g/l | 35 ml/100 l water | 14 days | 0.02 |
| Spinosad 480 g/l | 20 ml/100 l water | 3 days | 0.02 |
| *Bacillus thuringiensis* 32000 IU/mg | 100 g/100 l water | N/A | No MRL Required |

Mamay et al. (2014) reported that the infestation rate of Carob moth in Sanlıurfa (Turkey) pomegranate orchards was 45% in 2011 and 61% in 2012. Mirjalili and Poorazizi (2015) recommended an integrated method for the control of carob moth as follows: 1- prevention, 2- mechanical, and 3- biological, methods. They suggested that using of resistant cultivars is utmost important and thus using of giant fennel for keeping away the pest is very important. According to authors, collecting and burning remainders of old fruits from the previous year is important. For the biological control, they suggested the propagating and releasing Trichogramma wasps to the orchard.

## Leopard Moth

Leopard moth (Zeuzera pyrina) (Fig. 21), also called fruit borer, is one of nearly 700 species of Cossidae moths and may be the most damaging moth for pomegranates.

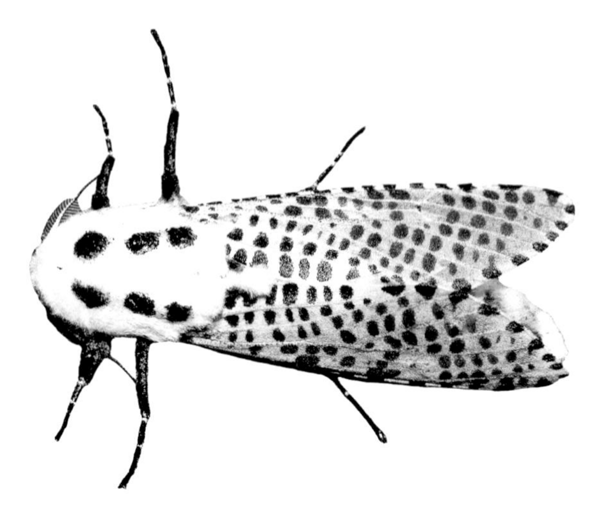

**Figure 21.** A view of a Leopard Moth

## Identification

The large white thorax of the Leopard moth displays 6 dark spots arranged in two lines (Fig. 20). The wings are almost transparent and covered in dark spots. The Male has comb-like antennae which is a significant distinguishing characteristic for the males and females. The Leopard moth lives on woods of almost every plant, preferably; apple, pear, plum, pomegranate, raspberry, olive, walnut, chestnut and carob. The forewing of the Leopard moth ranges between 2 and 6 cm depending on the geography and climate. Eggs are oval, flattened at both ends, smooth and yellowish pink in color. Larva has brown or shinning black head. Pupa is reddish brown in color.

## Life Cycle

The Leopard moth spend two or three winters in its larval form and feeds on the stems and leaves of various trees and shrubs. The larvae requires nearly two years to complete its growth before emerging to pupate. Pupa period is generally during the spring time from May to June. After that, adults may appear and they generally fly at night and rest on tree trunks by day (UC IPM 2015).

## Damages

Adults lay their eggs onto fruit barks. Larvae bore into the wood of branches and sometimes the main trunk (Fig. 22). If branches are severely attacked, they are likely to snap off in high winds. Sometimes, they cause the entire tree to die and fall.

**Figure 22.** A view of damage on pomegranate tree caused by Leopard Moth

**Management**

For a successful management of the Leopard moth, it is important to not establish orchards near to the host plants of the pest. Acer, citrus, vine, oak, ash, willow, maple, plane tree, olive, apple, cherry, pear, plum, and some similar trees are the host plants of the pest. Another important point is that, sometimes farmers use the woods of these plants as a support for the pomegranate trees when they are young. This type of application may bring the pest to the field. Therefore, one must pay attention to this situation. After the infestation by the pest of the field, the first step must be the cutting and burning of the damaged branches during larval stage. Thus, it may be necessary to spray insecticide during the adult pest generation (generally from June to August). Useful insecticides for the control of Leopard moth are given in the Table 6.

**Table 6.** Available active ingredients for the effective control of Leopard Moths

| Active ingredient | Suggested dose | Harvest interval | EU MRL (ppm) |
|---|---|---|---|
| Chlorpyrifos ethyl 480 g/l | 200 ml/100 l water | 14 days | 0.05 |
| Dimethoate 400 g/l | 75 ml/100 l water | 21 days | 0.02 |

## Mediterranean Fruit Fly

The Mediterranean Fruit Fly or Medfly (*Ceratitis capitata*) is among the most important agricultural pests in the world. It colonizes quickly unlike some other pests and can tolerate cooler climates; these characteristics make it a very important pest. The Medfly has spread throughout the Mediterranean region (where its name comes from), southern Europe, the Middle East, Australia and America. It has been reported to infest a wide range of agricultural crops like citrus, olives, fig, pomegranates, avocado, apple, almond, cherry, kiwi, papaya, peach, pear, plum, pepper, tomato, etc.

**Identification**

The adults of Medfly are slightly smaller than a house fly. They are about 3-6 mm in length. The color is yellowish with brown spots. The lower corners of the pest's face have white setae. Eyes are purple to reddish in color. The thorax is creamy white to yellow in color. Light areas of the body have very fine white bristles. Wings are the important parts for identification. The wings are mottled with distinct brown bands (Fig. 23). Female Medfly has an ovipositor whereas the males do not. Adult Medfly may live for 2-3 months and are generally found in foliage of trees. Their eggs are slender, smooth and curved, about 1 mm long. The eggs are shiny white in color. Larvae are white and elongate. The last instar is usually 7 to 9 mm in length. The pupa is about 4 mm long, cylindrical in shape, dark reddish brown in color and look-like a swollen grain of wheat.

**Figure 23.** A view of a Mediterranean Fruit Fly

**Life Cycle**

Medfly generally completes its life cycle in four stages: egg, larvae, pupa and adult in 3 weeks. However, development is temperature dependent. Eggs of the Mediterranean fruit fly are laid below the skin of the host fruit. They hatch within 2-4 days. However, this duration may be extended up to 16-18 days in cool weather. The larvae feed on leaves for about 5-10 days. Then, pupa stage generally occurs in the soil under the host plant. Thus, adults need about 5-10 days to emerge. Adults can live for up to 2-3 months. If temperature goes below 10°C, development ceases. Under cool conditions, the Medfly may require about 100 days or more to complete its life cycle (UC IPM 2015).

**Damages**

The damage of Medfly on the pomegranate is mainly caused by the oviposition in fruits and larvae feeding inside the fruits. However, decomposition of plant tissue by Medfly also invites other pests and may cause them to enter into fruits. Medfly generally affects cultivars which have thin peel, like Acco, Herskovitz, Mollar de Elche, etc. Some cultivars like Wonderful and Hicaznar have thick peels and Medfly does not prefer these cultivars. Pest could not lay eggs into the fruits with thick peel. Another important point for the Medfly damage is that cold weather decreases the activity of this pest. Therefore, pests generally do not prefer the late maturing pomegranates. Damaged young fruits may become distorted and usually fall from the tree. The larval tunnels provide entry points for bacteria and fungi and it may cause fruit to rot (Fig. 24).

**Figure 24.** A view of damage on fruits caused by Mediterranean Fruit Fly

**Management**

Control of the Medfly is very difficult because of its characteristics described above. Three control methods may be recommended currently: 1) attract-and-kill traps, 2) foliage baiting and 3) cover spraying. The use of attract-and-kill traps is a successful control measure for the Medfly (Fig. 25). The attract-and-kill trap consists of a pheromone and an insecticide. A pheromone is a chemical that attracts the males or females of Medfly. Substances attract insects and they enter the trap through three holes around the trap. Once the insects enter the trap, they try to fly towards the light and escape through the transparent top, during which the insects touch the insecticide, die and fall down into the trap. It is recommended to hang the traps on the southern parts of the pomegranate trees about 90 days before harvest. For the sensitive pomegranate cultivars, it was tested in Cyprus, where Medfly is a very destructive and abundant pest, and results show that 30 units $ha^{-1}$ of traps are sufficient for the effective control of Medfly (Kahramanoğlu and Usanmaz 2013).

**Figure 25.** A view of an attract-and-kill trap used for the control of Mediterranean Fruit Fly

### Cultural Control

One of the most important cultural methods for the controlling Medfly is field sanitation by the cleaning of infested fruits. Field sanitation may help to eliminate the source of eggs, larvae and pupas.

### Chemical Control

Medfly requires only a few minutes to lay eggs and thus chemical residues cannot kill adults within this short time. Therefore, chemical sprays often have not been effective in protecting fruit from medflies. For this reason, protein attractants together with an insecticide spray is recommended for the controlling of adults. This is called foliage baits. The female Medfly requires a source of protein for the maturation of eggs. This is why the Medfly feeds on the fruits to get the required protein. Therefore, baits encourage the adults (especially females) to feed on the protein, and the insecticides kill them. Bait-insecticide sprays must be used in combination with good sanitation practices to increase efficiency.

Application of bait (source of protein and insecticide) is not needed to be sprayed to the entire tree. Application of the bait is performed onto the southern parts of the trees on about 1-2 $m^2$ area of each tree on a row. One row is sprayed whereas the next row is not sprayed. The application is recommended to be performed about 2 m away from the soil, rather than the extreme top or bottom of the tree. Ripe fruits must not be sprayed. The first application is recommended to be performed when the first Medfly is observed within the orchard and if the weather conditions are suitable for the pest; i.e. >15°C. Depending on the population, weekly or twice-weekly re-applications are recommended. The re-application must be performed on the un-sprayed rows. During the applications, it is therefore of paramount importance to collect and destroy the fallen and unwanted crops from the orchard. Application of the bait is recommended to be performed in the morning. Midday and evening applications may have a risk of burning or scarring leaves and fruit. In some cases, if the population is too high, all fruit trees may be required to be baited. Rain is another important point for bait applications where heavy rains may wash off foliage baits and re-spraying is needed.

**Table 7.** Available active ingredients for the effective control of Mediterranean Fruit Fly as foliage baits

| Active ingredient | Suggested dose | Harvest interval | EU MRL (ppm) |
|---|---|---|---|
| Malathion 25% (insecticide) | 400 g/10 l water | 14 days | 0.02 |
| Nu-Lure (Protein) | 200 ml (40%)/10 l water | N/A | No MRL Required |

Another control method for the Medfly is cover spraying (Table 8). It differs from foliage baiting as it may kill all Medfly stages from egg to adult, but it may also kill any other pests of natural enemies. However, on the other hand, cover spraying does not provide an effective control of the Medfly out from the orchard and if the surrounding population is high, Medfly will continue to infest orchard. Thus, foliage baits would provide additional control.

**Table 8.** Available active ingredients for the effective control of Mediterranean Fruit Fly as cover spraying

| Active ingredient | Suggested dose | Harvest interval | EU MRL (ppm) |
|---|---|---|---|
| Dimethoate 400 g/l | 75 ml/100 l water | 14 days | 0.02 |
| Fenthion 525 g/l* | 150 g/100 l water | 14 days | 0.01 |
| Trichlorfon 80%* | 125 ml/100 l water | 7 days | 0.01 |
| Cypermethrin 250 g/l° | 30 ml/100 l water | 21 days | 0.05 |

* If needed, second applications must be performed with half doses.
° It is effectice but has low mobility within the fruit and has a residue risk for the fruits.

Several studies reported that spinosad bait treatments under laboratory conditions are effective in causing high mortality of Medfly (Raga and Sato 2005, Manrakhan et al. 2013). But, both studies were conducted by the combination of spinosad with protein hydrolysate under laboratory conditions. Kahramanoğlu and Usanmaz (2013) conducted a study in orchards and they reported low efficiency for spinosad. They suggest weekly application for this biological pesticide to have high efficiency on the control of Medfly.

## Mealy Bug

The citrus mealy bug, *Planococcus citri* (Risso) (Hemiptera: Pseudococcidae) attacks many host plants including pomegranates.

### Identification

Citrus mealy bugs prefer humid conditions and shape areas of plants. Adult mealy bugs' size ranges from 3 to 5 mm in length. The females are white to light brown in color and wingless. The bodies of adult females are coated with white wax which protects them from insecticides and make them an important pest for the crops. Since female mealy bugs are wingless, they need a vector to be transported to subsequent host plants. Females may live up to one month depending on the host plant. Females may lay 300 to 600 eggs. Males are similar in color to females. Adult males are winged differing from the females. The eggs are pink in color and hatch in 2 to 10 days.

### Life Cycle

Citrus mealy bugs pass through two additional instars before becoming an adult female or forming a male pupa. Each instar requires 1 to 2 weeks to complete. Female nymphs resemble adult female in appearance. However, male nymphs are more elongated than female nymphs. Male mealy bugs emerge from their pupa in 1 to 2 weeks. Males are generally about 4.5 mm in length. Under optimal conditions, a mealy bug mature from an egg to an adult stage in about 30 days. Citrus mealy bugs spend the winter generally as eggs on the upper soil, trunk, below leaves or lower branches of the tree. These eggs hatch generally in April and the crawlers move to green parts of the trees and fruits. Subsequent generations develop firstly on the fruit (Kerns et al. 2004).

**Damages**

Citrus mealy bug generally prefers the shade parts of the trees. If the fruits are not thinned and touching each other, adults hide there and protect themselves from enemies and insecticides. They feed on fruits and their infestation results in wilted and yellowed leaves (Fig. 26). It may cause premature leaf drops too. Feeding by the pest causes sugary honeydew occurrence on the fruits, resulting in the growth of sooty mold. Both spoil the appearance of fruits and also degrade fruit quality by reducing photosynthesis.

**Figure 26.** A view of damage on fruits caused by Mealy Bugs

**Management**

Management of the Citrus mealy bug needs an integrated approach. Only chemical applications can succeed where pests hide under leaves or crowded fruits. Thus, contact insecticides may not reach the pests and they will continue to survive. Summer rains may cause mortality by washing the mealy bugs from plants.

### Biological Control

Some studies reported that several natural enemies are effective at controlling citrus mealy bug. According to some literature, like (Kerns et al. 2004), biological control is the most effective method to control citrus mealy bug. They reported that, there are several parasitic wasps that prey on mealy bugs, and *Anagraphus* sp. is the most prevalent and important. In addition to parasitoids, some predators like; lady beetles, predaceous mites, lacewings, and syrphid flies can aid in mealy bug control.

### Cultural Control

As described in life cycle, adults hide in crevices between the foliage and the fruit. Thus, proper pruning to prevent shaping and most importantly proper thinning of the pomegranate fruits is very important in controlling mealy bugs. Performing fruit thinning by removing the fruits which are touching each other and leaving only one fruit on each spur is a very effective tool in controlling pests. Cleaning of equipment and harvest materials are useful for reducing the spread of this insect too.

### Chemical Control

Chemical control of this pest is often an inefficient method because of the nature of these pests where it hides in crevices between foliage and fruit. Some active ingredients are known to be effective against mealy bugs, but they are all contact insecticides, except Spirotetramat, and it is not recommended to use these insecticides. Chlorpyrifos ethyl is one of the contact insecticides which are still commonly being used. Thorough coverage is needed for this insecticide to be effective. Spirotetramat is known to be most effective insecticide against mealy bugs. But, it is highly recommended to be used in combination with proper fruit thinning. Application of the insecticides is recommended to be performed about 2 months before fruit maturing when the first infestations determined (Table 9).

**Table 9.** Available active ingredients for the effective control of Mealy Bugs

| Active ingredient | Suggested dose | Harvest interval | EU MRL (ppm) |
|---|---|---|---|
| Chlorpyrifos ethyl 480 g/l | 200 ml/100 l water | 14 days | 0.05 |
| Spirotetramat 100 g/l | 100 ml/100 l water | 28 days | 0.10 |

Ahmed and Abd-Rabou (2010) reported that *P. citri* is found to infest 65 plant species belonging to 36 families in Egypt and pomegranate is one of these species. Su and Wang (1988) highlighted that *P. citri* prefer to host clusters of fruit and inner canopy fruits. On the other hand, after development *P. citri* has protective wax secretions and these are the reasons that chemical control of this pest is often an inefficient management strategy, especially with contact pesticides. Similar findings for the inefficiency of chemical treatments have been reported (Tandon and Lal 1980, Ishaq et al. 2004). Kahramanoğlu and Usanmaz (2013) reported that Spirotetramat is a systematic pesticide and have higher effects than the chlorpyrifos-ethyl. Kerns et al. (2004) reported that chlorpyrifos-ethyl is commonly used for control of *P. citri* in Arizona. But, they also reported that it is a contact pesticide and thorough coverage is need for chlorpyrifos to be most effective.

## Mites

Citrus flat mites (*Brevipalpus lewisi*) are also damage pomegranates. They are very small pests and it is difficult to see them with the naked eye or even with a hand lens.

### Identification

They are translucent, flat and oblong. Adults have internal spots of red and brown. These mites spend the winter under flakes of bark on large branches. They move to the leaves and fruit in summer. The mites are red and flat, and the adults are about 0.25 mm long. They have 2 pairs of short legs at the front of the body and 2 pairs of short legs flanking the narrow abdomen (UC IPM 2015).

### Life Cycle

Citrus flat mite populations are low but detectable from late May until early June, when mean temperatures were lower than 20°C. Thus, with the increasing temperatures, mite populations begin to increase. Thus, in late September mite numbers begin to decline, along with declining temperatures. These periods keep changing depending on the climatic conditions of the area.

### Damages

Citrus flat mite feeding results in a brown scabbing or leathering of the fruit looks like sunburn. However, sunburn and other unknown causes of leathering can be mistaken for flat mite damage. The damage caused by flat mites also causes damage on the fruit surface next to the stem; otherwise it is not flat mite damage. Pest population starts increasing in June, peaking in July and August, and then gradually declines.

### Management

Predaceous mites may keep flat mites below economic levels. However, excessive populations need to be treated with insecticides to reduce population and prevent damages. Sulphur application is the best control method to prevent flat mite damage.

## Thrips

Some thrips may occur on pomegranates, but significant damage is not observed.

### Identification

Thrips (*Scirtothrips dorsalis*) generally prefer feeding on new growth of plants. This species is yellowish and greenish in color and have two black stripes on the body. Eggs are dirty white in color and bean shaped. Newly hatched nymphs from the eggs are reddish in color and turn yellowish brown as they grow.

### Life Cycle

Females lay on an average 50 eggs on the under surface of leaves. The incubation period is 3-7 days. Pupa period lasts for about 2-5 days.

### Damages

Nymphs and adults lacerate and suck the contents of buds, flowers, leaves and fruits. When thrips feed on leaves, leaf tips turn brown and get curled. Sucking the cell-sap of the flowers by thrips may cause drying and shedding of flowers. Scrapping on fruits leads to scab formation, reducing market value (Fig. 27). Damages on the young fruits by thrips may cause fruits to become susceptible to cracking.

### Management

Removing and destroying all infected plant parts is essential in effective controlling. Following insecticides are known to be affective for the control of thrips.

**Figure 27.** A view of damage on fruits caused by Thrips

**Table 10.** Available active ingredients for the effective control of Thrips

| Active ingredient | Suggested dose | Harvest interval | EU MRL (ppm) |
|---|---|---|---|
| Acetamiprid 20% | 40 ml/100 l water | 14 days | 0.01 |
| Imidacloprid 200 g/l | 20 ml/100 l water | 14 days | 1.00 |

## Whitefly

Whiteflies are tiny, sap-sucking insects which may damage pomegranate fruits and leaves, but the damages are very rare. They excrete sticky honeydew, like aphids and cause yellowing or death of leaves.

### Identification

Their name comes from the mealy white wax covering the adult's wings and body. Adults are tiny insects with yellowish bodies and whitish wings. The body of whiteflies has a triangular shape when seen from above. Eggs are white in color for the first two days and later turn into black. Pupae are oval in shape and white in color.

### Life Cycle

Whiteflies usually occur as a colonized group on the undersides of leaves. Whiteflies develop rapidly in warm weather and populations grow rapidly if there are no natural enemies. Most common whiteflies are: greenhouse whitefly (*Trialeurodes vaporariorum*) and potato whitefly (*Bemisia tabaci*). Whiteflies generally lay their eggs on the undersides of leaves. The eggs hatch and grow through four nymphal stages. At the end of the nymph stage, the legs and antennae are reduced and older nymphs become stationery. The winged adult emerges after the forth nymph stage. Duration of egg period is about one week where the duration of the coming four nymphal stages are: 6, 2, 3 and 4 days, respectively. Pupae stages go on for 3 weeks and life span of adults may be up to 40 days. Female whiteflies start laying eggs about 2 days after emerging. All stages of the pest feed by sucking plant juices from leaves, flowers, fruits, etc. Development and egg laying ceases below 7-8°C. Temperatures above 30°C are not favorable for whitefly development.

**Damages**

Whiteflies feed on phloem sap and excrete honeydew, like aphids (Fig. 28). Large populations can cause leaves to dry and fall off. Some whiteflies may transmit viruses to pomegranates. As described above, whiteflies are not common problem for pomegranates, but higher infestations can be a problem and need to be treated.

**Figure 28.** A view of damage on fruits caused by Whiteflies

**Management**

Since the whitefly damages are not occasional, sometimes special management is not necessary. But, heavy infestations are not easy to manage.

### Biological Control

There are some natural enemies for the whiteflies. Some of them are; lacewings, lady beetles and minute pirate bugs. For this reason, conscious application of any insecticides is of paramount importance to not kill or damage the natural enemies. This is a very important aspect of whitefly management. Control of dust and ants may play an important role in controlling whiteflies in which they protect pests from natural enemies.

### Cultural Control

During the non-mobile nymph and pupa stages, hand-removal of leaves and eradication may be an effective control measure. Reflective plastic mulches can repel whiteflies, especially away from small plants. Yellow sticky traps can be hanged inside the orchard both for population measurement and possible control. Since whiteflies do not fly far, many traps are required for effective control.

### Chemical Control

Many insecticides have been tested against whiteflies, but none of them have had effective control on the heavy populations. Most of the insecticides kill only those whiteflies which come in direct contact. For this reason, application of the insecticides must be performed carefully to thoroughly cover the tree, including undersides of all infested leaves. Whiteflies may easily build up resistance against the insecticides, if they survive. For this reason, it is not recommended to use insecticides. However, in any case, following insecticide may be effective, if used correctly.

**Table 11.** Available active ingredients for the effective control of Whiteflies

| Active ingredient | Suggested dose | Harvest interval | EU MRL (ppm) |
|---|---|---|---|
| Imidacloprid 200 g/l | 20 ml/100 l water | 14 days | 1.00 |

CHAPTER 6

# POMEGRANATE DISEASES

## Fruit Spots

The causes of fruit spots may be *Alternaria alternate* (Alternata fruit spot) or *Cercospora punicae* (Cercospora fruit spot). *C. punicae* also may cause similar spots on leaves.

### Symptoms

Alternaria fruit spot occurs as small reddish brown circular spots fruits (Fig. 29). Increase in the infestation coalesces to form larger patches and the fruits rot. Cercospora fruit spot occurs as light brown spots on fruits (and as indicated above same spots may appear on leaves). Spots on leaves may be black and elliptical. Affected areas on the leaves become flattened and depressed with raised edge. Infestation on leaves may cause it to dry up.

**Figure 29.** A view of an Alternaria fruit spot damage on fruits

### Cultural Control

Primary source of disease is infected plant parts. For this reason, removing and destroying of the infested plant tissues is highly important for the control of fruit spots.

Secondary source of inoculums is wind born conidia.

### Chemical Control

Application of the following (Table 12) fungicides is effective in controlling fruit spot disease. Application is recommended to be performed when first symptoms appear on fruits.

**Table 12.** Available active ingredients for the effective control of Alternaria fruit spot

| Active ingredient | Suggested dose | Harvest interval | EU MRL (ppm) |
|---|---|---|---|
| Mancozeb (80%) | 200 g / 100 l water | 14 days | 0.05 |
| Propiconazole (15%) + Difenoconazole (15%) | 50 g / 100 l water | 7 days | 0.05 & 0.10 |

## Heart Rot

Perhaps the most important problem of the pomegranate is the Heart rot (also known as: Black heart). Many studies have been conducted to determine the causing agent of heart rot (Barkai-Golan 2001, Crites 2004, Michailides et al. 2011, Zhang and McCarthy 2012, Ezra et al. 2014, Kahramanoğlu et al. 2014). The fungus *Alternaria* spp. was isolated as the major source of black heart disease. This fungus enters the blossom and then grows in the resulting fruit. The infection process is not completely understood, and the type of *Alternaria* resulting in the infection is still being isolated.

### Damages

Transmission of *Alternaria* spp. to the heart of pomegranate fruit happens during bloom and no infection may occur after fruit set. The fungus causes decay of arils of pomegranate fruits ranging from sections to all the arils, without obvious external symptoms except slightly abnormal skin color or changes in shape (Fig. 30). The fungus grows inside the fruit and causes fruit rot (Fig. 31). When a pomegranate has heart rot, it is no longer marketable and the producer risks losing crop income. However, *Aspergillus niger* and *Penicillum* sp. may cause same or similar damages.

### Management

Removal of old fruit from the tree is suggested to eliminate the potential source of the fungus. Since the disease is being transmitted to the heart of the fruits during bloom, covering the trees with a 125 μ net eliminates the infection of fruits with *Alternaria* spp. But this practice results in a reduction in the number of fruits and it is not practically acceptable. Transmission of fungus might be by miscellaneous insects (bigger than 120 μ) instead

**Figure 30.** An outside view of a fruit with Heart Rot damage inside it

**Figure 31.** A view of Heart Rot damage

of wind and rain and this pest presumably could be various bugs of the genus Leptoglossus. So, controlling of the bugs during blooming period helps to control this disease (Nizam et al. 2015).

For the effective control of the pathogen, application of copper based formulations is suggested to be used immediately after pruning during autumn period is recommended. Aplication of the one of the following (Table 13) fungicides at least twice during blooming period could be an effective measure to control Heart Rot infestation. First application should be performed when flowering starts and approximately 30% of the flowers are fully-open.

**Table 13.** Available active ingredients for the effective control of Heart Rot

| Active ingredient | Suggested dose | Harvest interval | EU MRL (ppm) |
|---|---|---|---|
| Boscalid (25%) + Pyraclostrobin (12%) | 50 g/100 l water | 7 days | 0.05 & 0.02 |
| Propiconazole (15%) + Difenoconazole (15%) | 50 g/100 l water | 7 days | 0.05 & 0.10 |
| Azoxystrobin (25%) | 60 g/100 l water | 7 days | 0.05 |

## Bacterial Blight

*Xanthomonas axonopodis* pv. punicae. is the main cause of bacterial blight on pomegranate fruits and leaves. The pathogen spend the winters in infected leaves and increases through years.

### Symptoms

Bacterial blight causes dark colored irregular shaped spots on leaves and on fruits (Fig. 32). The pathogen may also infect stem and branches and cause cracking symptoms. Spots on fruits are dark brown in color. Continuous rainfall, temperatures between 20-35°C and relative humidity between 60-90% are favorable conditions for the development and spread of the disease.

**Figure 32.** A view of Bacterial Blight damage

### Cultural Control

Wide row spacing is an important cultural control measures. It produces good air flow and reduces relative humidity and helps to reduce pathogen infestation. Thus, removing and destroying of affected branches and fruits are crucial for the control of Bacterial Blight.

### Chemical Control

For the control of the pathogen, Bordeaux mixture application immediately after pruning during autumn period is recommended. After 10-15 days interval, second application with copper based formulations is suggested for the effective control.

## Pomegranate Wilt

Pomegranate wilt (*Ceratocystis fimbriata, Fusarium oxysporum* and/or *Rhizoctonia solani*) is another important pathogen which affects the whole tree and causes tree losses.

### Symptoms

Leaves of the affected plants start yellowing followed by drying and drooping of leaves. This causes the entire tree to die in the same year. If the infected tree is cut near to root as open cross-section, dark greyish-brown coloration of wood is seen (Fig. 33). Soil is the primary source of pathogens. Disease is more in heavy soil and infestation increases with the increasing soil moisture.

**Figure 33.** A view of Pomegranate Wilt damage

## Cultural Control

Since soil is the primary source of pathogen, soil solarization or sterilization before planting would be helpful in preventing disease infestation. On the other hand, spaced planting of trees (5 m × 3 m) is also a prevention measure for the pathogen. Dense planting enhances pathogen infestation.

## Chemical Control

Application of following pesticides immediately after the occurrence of symptoms is recommended. The highly infested trees must be removed and destroyed. At initial stage, it is advisable to drench with 2 ml Propiconazole + 4 ml Chlorpyrifos ethyl per litre water. However, it should be kept in mind that Chlorpyrifos ethyl may damage flowers and reduce yield. Therefore, in case of the selection of this pesticide, it is recommended to wait until the end of flowering.

**Table 14.** Available active ingredients for the effective control of Pomegranate Wilt

| Active ingredient | Suggested dose | Harvest interval | EU MRL (ppm) |
|---|---|---|---|
| Propiconazole (15%) | 50 g/100 l water | 7 days | 0.05 |

# CHAPTER 7

# WEED MANAGEMENT

Weeds are defied as "the plants growing out of places, where they are not desired and their disadvantages are more than their advantages". Many weeds compete strongly with pomegranate trees depending on the climate and geographical region. They compete with pomegranates for nutrient, water and sun light (generally for newly planted trees). For this reason, weed control is very important for economic production of pomegranates. However, on the other hand, weeds are also hosts for natural enemies and other biological elements of the earth. According to E. J. Salisbury (1961): "everyone (except evils) have some admirable qualities: a thief can be a compassionate father and husband in his home; similarly a very harmful weed can be a pleasant flower in another place". Therefore, the aim of weed management should be to maintain the population below economic threshold rather than eradicating all. Before weed management, it is highly important to remember following features of weeds:

- Weeds (and buds) require a waiting period for internal maturity and external environmental conditions for being ready to germinate. This waiting period is called as dormancy. Most of the cultivated crops do not have dormancy and this characteristic makes weeds a troublesome problem for farmers. Some weed seeds (i.e., Wild mustard; *Sinapis arvensis*) can wait 35 years in soil to germinate. Therefore, controlling of weeds before seed production is highly important.
- Weather conditions and water availability plays a very important role in breaking dormancy. This knowledge can be used to control weeds. For example, drip irrigation is highly suggested where the water reaches only the weeds near the drip lines and other seeds do not germinate.
- Immediately after germination, weeds have a period called as "dependent vegetative period" where plants take the required energy from their storage organs (seeds, rhizomes) until the beginning of photosynthesis. During this period, weeds are more susceptible to environmental conditions where they can not produce their own foods.

- Another important characteristic of weeds is that they can have ability to reproduce vegetatively. This characteristic is generally seen on perennial weeds and it is difficult to control these weeds. Knowing these features is very important for the control of weeds, where wrong practices may cause an increase in the population. Followings are the examples for vegetative reproduction abilities.
    - **Stolon.** Some weeds reproduce with its stolons and other horizontal plant parts: These plant parts horizontally grow on soil and thus produce roots and new plants. (i.e., Bermudagrass; *Cynodon dactylon* Crabgrass; *Digitaria sanguinalis* (Fig. 34))

**Figure 34.** Photo of *Digitaria sanguinalis*

    - **Rhizoms.** Some weeds have rhizomes under soil and they produce roots to develop new plants (i.e., Johnsongrass; *Sorghum halepense* (Fig. 35), Couch grass; *Agropyron repens*)

**Figure 35.** Photo of *Sorghum halepense*

- **Tubers.** Tubers are the more developed rhizome tips. These plant parts are generally used for storage organs. (i.e., Jerusalem artichoke; *Helianthus tuberosus*, Purple nutsedge; *Cyperus rotundus* (Fig. 36))

**Figure 36.** Photo of *Cyperus rotundus*

- **Roots.** Many weed species can produce new shoots from their horizontal roots (Bugwoodwiki; *Cirsium arvensis*, Field bindweed; *Convolvulus arvensis* (Fig. 37))

**Figure 37.** Photo of *Convolvulus arvensis*

- **By breaking.** Some parts (leaves, stem, etc.) of some weeds can easily produce roots and shoots if cut and they fall down onto soil (Purslane; *Portulaca oleraceae* (Fig. 38))

**Figure 38.** Photo of *Portulaca oleraceae*

- Weed seeds can easily disperse to different environments in time by wind, water, soil, animals, animal manures, equipment and human beings. Thus animal manures are recommended to be composted before use to eradicate the weed seeds. Composting duration depends on the type of manure and the season
- Weeds are the hosts and source of aphids. Before the spring flushes, weeds must be cleaned from the orchard and surroundings to reduce or eliminate aphid infestation.

## Management

Management of the weeds needs an integrated approach for economic, viable and environmental success. For this reason; integrated management of weeds needs to include cultural, mechanical, physical and biological methods together with chemical applications.

## Cultural Control

Correct adjustment of planting distance is the first step for the cultural control of weeds. The detailed explanations have been done in the orchard establishment section of this book. The best distance for the average growing pomegranate cultivars: Wonderful, Hicaznar, Mollar de Elche and etc., is 3 m between plants and 5 m between rows. Apart from planting distance, irrigation method is very important. Most of the weed seeds need water to germinate, so drip irrigation method or micro sprinkler lines are recommended where the weeds only germinate around the drip lines. Last but not the least, prevention of infection is very important. For this reason, irrigation water must be filtered to clean weed seeds. Animal manures must be composted before use and all agricultural machines/ equipment must be cleaned before the introduction into the orchard.

## Mechanical Control

Soil management always needs to be performed before weeds produce seeds. Time for soil management needs to be arranged at the times when the climate and soil is not suitable for vegetative reproduction. Recommended time for mechanical control of weeds is before the spring flush appears.

For the management of perennial weeds, it is important to plough the soil. However, this can only be done before planting, not in the established orchards. Thus, the weed parts under the soil comes up and it is required to collect them by rake and taken away. Disk harrow and router breaks the rhizoms, stolons and bulbs and breaks the apical dominance. Thus, the numbers of weeds increase! Therefore, cultivator is more effective to use in such weed population (Fig. 39).

**Figure 39.** A view of a cultivated soil

Soil management needs to be performed during the seasons when the storage organs of weeds (stolon, rhizome, bulb, etc.) are not active and sometimes need to be repeated more than one time to clean weeds. For the second control measure, brushers or moving may also be used which is very effective against newly emerged weeds and does not reproduce perennial weeds. Moving does not damage soil and it may be repeated regularly to control weed populations depending on the economics. Soil management machines should be selected according to the weed species and characteristics.

## Physical Control

Solarization and mulching are the effective control methods for weeds. Solarization can be performed before orchard establishment, not after. For solarization, transparent/white materials are used and thus the energy of sunlight heats the soil. Therefore, this causes the weed seeds, diseases and pests to be eliminated from soil. Mulching is a process of covering the soil surface with any material. The goal for mulching is to prevent sunlight penetration into soil. Mulching is beneficial where it:

- prevents water loss from soil,
- prevents soil against frost,
- prevents soil borne diseases (i.e., *Alternaria*),
- prevents leaching,
- protects the physical structure of soil,
- increase the amount of beneficial nutrients (live mulch),
- prevents the contact of crop stem to the soil and thus prevents rotting,
- increase the soil temperature and accelerate plan growth

Black plastic, newspaper, wood chips, peat moss, straw and wood shaving may be used as a mulch material. Plants especially from the Fabaceae family may be used as cover crops (live mulch). It is also beneficial because it:

- controls soil erosion,
- increases the nitrogen content of the soil and balances water,
- protects soil organism, regulates soil aeration and balances soil temperature,
- does not affect physical structure of soil and protects the roots of crops,
- prevents some disease infections,
- controls nematodes

## Biological Control

Biological management is carried by using some living things which negatively affect weed populations and do not affect crops. These living things generally reduce weed populations below the economic thresholds, instead of eliminating them. Most of these living things are insects and disease agents. However, biological control by insects and diseases are weed specific and can not control all weed populations. For the biological weed management in pomegranate orchards (not young!), chickens, geese and sheep may be used during winter period when the trees shed their leaves.

## Chemical Control

Chemical control is easy to carry and gives quick results. However, it negatively affects the environment and in long term may cause weeds to become resistant to herbicides. Therefore, chemical control methods need to be carried correctly and they must be the LAST choice of farmers. Success of the herbicide application depends on:

- Correct application timing,
- Correct application dosage,
- Correct application equipment (hand pulverizator and atomizer CAN NOT be used),
- Correct application pressure,
- Herbicides cannot be mixed with any pesticides or fertilizers,
- Water usage: 20-50 lt/da (depending on the herbicide),
- Application must NOT be done under rainy, windy and very hot climatic conditions,
- Herbicides should NOT touch crops, especially total herbicides,

- Soil herbicides must be mixed into soil after application,
- It is recommended to use micronherbi for the systematic leaf herbicides, i.e., Glyphosate.

Glyphosate is a systemic herbicide and can be applied throughout the season but special care should be taken to avoid applications on windy days and they should not be applied on the pomegranate trees or base suckers of the trees. Sucker removal is important otherwise glyphosate can be translocated into the trees and will kill them (Fig. 40).

**Figure 40.** A view of Glyphosate damage on the suckers

Not only for Glyphosate but for all herbicides, special care should be taken. On the other hand, herbicide rotation is highly suggested for the prevention of weed resistance against herbicides. Annual weeds are controlled mainly with pre-emergence weed killers (herbicides) such as Oxyfluorfen and Pendimethalin. After germination, weeds are controlled by Glyphosate. Oxyfluorfen may also be used for contact killing of some weeds.

**Table 15.** Available active ingredients for the effective control of Weeds

| Active ingredient | Suggested dose | Harvest interval | EU MRL (ppm) |
|---|---|---|---|
| Glyphosate 480 g/l | 300 ml/da for annual weeds<br>600 ml/da for perennial weeds | 21 days | 0.10 |
| Pendimethalin 450 g/l | 500 ml/da | 21 days | 0.05 |
| Oxyfluorfen 240 g/l | 250 ml/da | 21 days | 0.05 |

# CHAPTER 8

# PHYSIOLOGICAL DISORDERS

## Fruit Cracking

Fruit cracking is a serious problem of pomegranate (Fig. 41). All cracked fruits lose their value for fresh market. These fruits may only be used for concentrated juice production, if they are not affected by fungus, bacteria or insects. For 100% natural fresh squeezed juice production, where the world trend is for that kind of juice, cracked fruits are not preferred to use.

**Figure 41.** A view of a fruit cracking

Cracked fruits are also susceptible to storage disease. There are many causes of fruit cracking including:

- The main cause of fruit cracking is irregular and/or heavy irrigation or rain after long dry period.
- The extreme variations in day and night temperatures may be the other reason of fruit cracking. Temperature may also affect permeability of the cell walls and bio-chemical processes of the cells and enhance cracking.
- Occurrence of hot wind during summer months also enhance fruit cracking by passes causing water loss on fruit surface and making peel hard and inelastic.

- Another important reason of fruit cracking is the deficiency of boron and calcium.
- Unbalanced fertilization may also cause cracking if the inside of fruit is developing rapidly than the peel.
- Excessive application of nitrogen, especially before maturing may cause fruit cracking.
- Severe attack of thrips, mites and some other pests may also enhance fruit cracking where these pests damage the peel and make it sensitive against cracking.

**Management**

Cracking management starts with the site selection for the pomegranates. The appropriate site is recommended to have little or no rain incident before and during harvesting period. For a successful control of cracking, the main important step is regular irrigation. General irrigation requirement of pomegranate trees is given in irrigation section of present book. The amounts may have slightly changed according to cultivars. However, the most important point here is that the monthly amounts are also changing depending on the climate, geographical region, the age of tree and the yield on the tree. Therefore, irrigation of the plants from fruit setting to maturity with adequate quantity of water at regular intervals is very important. Irrigation of the pomegranates during harvest is also very important. Some growers stop irrigation before harvest which damages tree physiology and stimulates fruit cracking. Conservation of soil moisture during very hot months is very important due to reducing rapid evapotranspiration. Mulching may be performed for this reason.

Planting of suitable wind break around the orchard at a right angle to the direction of prevailing wind is helpful for the prevention of cracking. Two applications of boron (5%) (2 g/liter) one during fruit enlargement stage and one before harvesting are helpful for the prevention of fruit cracking. Two applications of calcium (calcium nitrate "10%" 2-4 g/liter) firstly in full blossoming and secondly one month post full blossoming are also recommended. Spraying of gibberellic acid (40ppm) on the young fruits also minimizes the incidence of fruit cracking.

## Sun Burn

Sun burn is also an occasional problem for pomegranates. This disorder generally occurs on young trees, but false pruning may also enhance sun burn on trees. Surface skin of pomegranate fruits facing sun turns brown or bronze color and may also become dark brown due to scorching (Fig. 42). High temperature along with excessive light, drought, and low relative humidity is usually responsible for sun scald injuries. Hot winds is the another important cause of sun burn.

**Figure 42.** A view of a fruit with sun burn damage

**Management**

The first step for the control of sun burn is to avoiding heavy pruning and developing a good canopy by proper pruning and plant nutrition to provide shade to fruits. Otherwise, or on young fruits, protecting of the fruits from direct sunlight by bagging or covering with newspaper or similar materials is recommended. Approximately one and half month before harvesting, covering materials must be collected to ensure proper coloring of the pomegranates.

Spraying kaolin thrice at 15 days interval during hot summer months, is another effective way of controlling sun burn. However, this method is somehow expensive. First spray should be performed just before hot weather conditions over whole canopy and fruits with of 5%. The next two, and sometimes three, sprays may be 2.5% of kaolin. If heavy rain washes the kaolin, spray interval must be reduced.

## Shape Disorders

Pomegranate fruits are globose in shape, i.e., flattened at the poles. Pomegranate fruits are sensitive to shape disorders. The reason of abnormal shapes may be mechanical damages during fruit set and/or thrips damage during flowering and fruit set. Damages during the fruit set cause fruits to be abnormal in shape (Fig. 43). Sometimes, inadequate pollination may be the reason of the shape disorders. For the control of shape disorders, it is recommended to control thrips and aphids infestation during flowering and fruit set. This will prevent damages on the peel of the fruits and may have a control over the abnormal shapes. Thus, during fruit thinning, it is highly recommended to start fruit thinning from the abnormal fruits.

**Figure 43.** A view of a fruit with abnormal fruit shape

## Hail Damage

Hail damage is not common in pomegranates. However, it is related with the climate and may cause extreme damages in some countries. Thus, the extent of losses varies from mild to high. Hail storms can vary in intensity and duration when they occur during the growing season. In addition to the damages on fruits, damages on the leaves and trunk may also be seen. Sensitive leaves of plants become shredded, pock marked or ripped by hail. Hail damage can completely kill young trees. Hail damage increases the susceptibility of fruits to decay. Fruit may be damaged and fall off to the ground.

### Management

Prevention of the damage by hail is only possible by the installation of nets over the trees which are not economical. Another important point is to avoid establishing orchards in the climatic conditions where hail damage is possible. On the other hand, after a hail storm it is important to know some practices which may help to restore trees and may protect fruits. Damaged branches are sensitive to fungus and bacteria where the damages present entry points for pathogens. First thing, after a hard hail storm, is pruning out the highly affected branches. After that, application of nitrogen fertilizer to the impacted plants is beneficial for the regrowth of the branches. 10% Bordeaux paste is recommended to apply to the cut ends of the branches. If the damage on the fruits is too heavy, thinning of the fruits may be performed to well manage the rest. One application of 2 g/l boric acid is beneficial to apply before harvest. Additionally to boric acid, 2 g/l zinc may also be applied to increase the efficiency. This may help to avoid rotting in fruits because of surface injury. Approximately 10 days after microelements application, application of 5 g/l urea is recommended to enhance regrowth of the branches.

## Mechanical Damage

Similar with many other fruits, pomegranate peel is sensitive to mechanical damages. Mechanical damages may be caused on fruits because of hard winds and incorrect pruning. Wind causes fruits to lurch on trees and touch the spiny parts of the branches (Fig. 44). Thus, mechanical damage may occur on fruits which may then enhance the growing of pathogens on or inside the fruits. Therefore, protection of the orchard from hard winds by wind breakers is very important to prevent mechanical damage. On the other hand, as described in pruning section, branches which may touch each other and cause crowding in trees should be removed during pruning to prevent mechanical damages.

**Figure 44.** A view of a fruit with mechanical damage

# CHAPTER 9

# FRUIT THINNING

Many trees tend to produce more fruits than they can carry. This is due to plants trying to guarantee their generations. Most trees produce many flowers and if all conditions, i.e. weather, fertilization, irrigation, etc., are right, majority of flowers on a tree may set fruit (Fig. 45). Every individual fruit needs carbohydrates to grow and in this case fruits compete with each other for carbohydrates. The carbohydrate source of each tree is particular and can not feed all fruits. Therefore, excessive number of fruit causes fruits to be smaller than the possible size. On the other hand, excessive number of fruits cause tree vigor to decline which then reduce the next season's yield too.

**Figure 45.** A view of an excessive number of fruit set as a cluster on a spur

For these reasons, fruit thinning has several benefits where the most important one is to increase fruit size and quality. Correct fruit thinning allows each fruit to develop to its possible maximum size (if other factors are right) and little or no reduction occurs in tree vigor. Also correct fruit thinning allows each fruit to receive more light which improves fruit color and flavor. Another benefit of fruit thinning is that, it prevents the breakage of branches. Last but not least, many pests and diseases damage those fruits which are touching each other, like mealy bug, so fruit thinning reduces chances of infestation.

Pomegranate trees can flower throughout the year depending upon the climatic conditions. The trees generally begin to bear fruit in the third year. Main flowering, which produces high quality fruits, occur during February to April in northern hemisphere. Fruit becomes ready for harvesting 5 to 7 months after the appearance of blossoms depending on the cultivar. Both self and cross-pollination is known in pomegranate. Pomegranate quality highly depends on the quantity of fruits on the tree. Leaving all fruits on the tree causes a considerable decline in the quality of fruits. Therefore, fruit thinning is highly recommended to increase quality of fruits. Pomegranate fruits mostly occur in clusters. The general rule of fruit thinning is to leave alone the terminal fruit, while removing the axillary fruits (Fig. 46).

**Figure 46.** A sample for fruit thinning

The best time for fruit thinning is after the fruit set when the fruits are about walnut size. This is because in some cases natural drop occur on pomegranate flowers and not every flower produces fruits. Early thinning may result in less fruit because of unexpected natural drop and thinning too late may reduce the chances to increase fruit size. Fruit thinning also gives a chance to remove damaged fruits by insect or disease. Preferred space between the fruits is 7 to 10 cm apart on each branch. Only 1 or 2 fruits should be left on small braches. Thinning should be done to allow a closer spacing near the base of the branch and a wider spacing near the tip of the branch to avoid the branch bending (Glozer and Feguson 2011).

Depending on the market demand and marketing strategy, the number of fruits per tree can be adjusted to reach highest profit. When the number of fruits increases, average weight and quality of the fruits decrease. By thinning out the fruits, the number of fruits is decreased however, the fruit weight and quality increase. Thus, the total yield generally becomes similar at both conditions. The important point here is the quality of fruits which in turn affects the profit too. But, one must remember that thinning out the fruits is also a cost for farmers. Jafari et al. (2014) conducted a study to determine the effects of severity of hand-thinning on fruit size and quality attributes of 'Malase Torshe Saveh' pomegranate cultivar. They reported that thinning out the fruits from 87 to 66, caused an increase in the average fruit weight from 292 g to 359 g. However no significant differences were determined on fruit yield which was 25.2 kg/tree on control plants and 23.3 kg/tree on thinned trees.

# CHAPTER 10

# HARVEST AND FRESH FRUIT PROCESSING

## Harvesting

Fruits become ready for harvesting about 5 to 7 months after the appearance of blossoms. They are harvested with special secateurs. Outer appearance, total soluble content of the fruit, juice content of the arils and inner color are the important characteristics need to be considered before harvesting. Due to the extended bloom period, fruit can mature at different stages. If the fruiting was well and fruit thinning was performed by leaving only fruits of first flowers, all fruits can be harvested at once. However, in other situations, fruits must be harvested when they mature; first of all, the apical fruits, which come from first flowers, then the axillar fruits which come from the second and third flowers and at last the remaining fruits. Full yields generally occur after $6^{th}$ year of planting. Yields in a very productive tree, depending on the cultivar can reach up to 60-100 kg.

Pomegranate fruits can not be matured after harvest. Therefore, it is very important to pick the fruits at correct time. However, on the other hand, delay in harvesting causes cracking in fruits. Fruits for fresh consumption must be picked by hand (by shears) and carefully handled. Bruise on skin may cause a dark blemish on the shiny rind, but not actually damage the inside of the fruit. However, the skin damages cause the inner quality to decrease in time during storage. Mechanical damage also increases moisture losses and reduce weight during storage. And, only the external appearance is enough for the reduction in the commercial value. Shears should be used to cut the fruit off. The fruit should be protected from sharp twigs (Fig. 47). The fruits are sensitive and should be placed into bags or boxes carefully.

Before harvesting, workers should wait until the dew dry on fruits. Otherwise, blemish occurs on fruits which cause damages during storage. The second important point is the protection of harvested fruits from direct sunlight. Immediately after harvest, fruits begin to lose weight through transpiration. The main cause of transpiration is the temperature and fruits must be kept at cooler temperatures and protected from sunlight. Harvested fruits may be stored under tree or shaded areas until

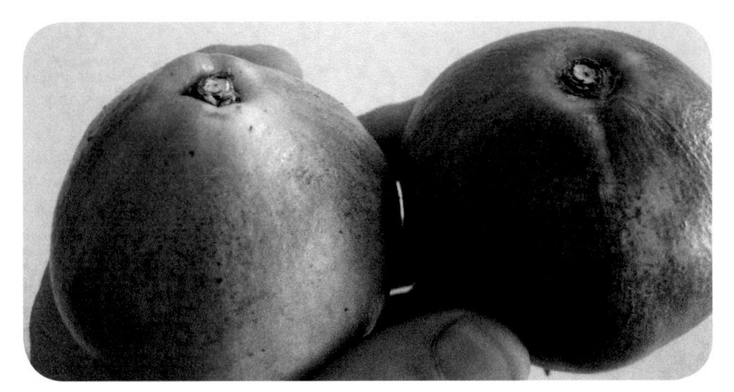

**Figure 47.** Correct harvesting of pomegranate fruits

they are transported to the factory for packing or should be packed at the field immediately and transported to markets or storage. In some cases, frequent transport to final destination is preferred to reduce weight loss. If possible, it is highly recommended to transport in controlled conditions, i.e. temperature and humidity.

## Quality Standards

Quality standards are very important for the exportation of the products. The reason of the development of quality standards is the increasing consciousness of consumers about health and food safety. Therefore, it is very important for the pomegranate farmers and traders to know the standards. There are two big bodies in the world determining standards for fruits and vegetables and their standards are being accepted in all over the world. One of them is United States Department of Agriculture and it does not have any standards on pomegranates yet. The other important body for the food standards is European Commission where the standards are prepared based on the Codex standards. According to EU standards, pomegranate fruits are classified as "Extra", "Class I" and "Class II". In all classes, subject to the special provisions for each class and the tolerances allowed, the pomegranates must have some minimum requirements and must be (CBI 2015):

- whole; (Fruits must not have any damage or injury spoiling the integrity of the produce. Slight defect in crown is permitted);
- sound; produce affected by rotting or deterioration such as to make it unfit for consumption is excluded; (Fruits must be free from disease. In some cases, fruits may be damaged by Heart rot or rotting without having external signals or very slight abnormal color but liable to make the produce unfit for consumption upon arrival at its destination, are to be excluded);

- clean, free of any visible foreign matter; (Fruits must be free of dust and chemical residue);
- free of pests and pest damages on the general appearance of the produce; (The presence of pests can detract from the commercial presentation and acceptance of the pomegranate);
- free of abnormal external moisture, excluding condensation following removal from cold storage; (This does not include moisture on produce following release from cool storage but applies to excessive moisture, i.e., free water lying inside the package);
- free of any foreign smell and/or taste; (in some cases, storing or transporting pomegranates with other fruits or products may cause them to absorb abnormal smells and/or tastes);
- free of damage caused by frost;
- free of damage caused by low and/or high temperatures;
- free of sun burns affecting the arils of the fruit (Sun burns might affect the arils color rather than outside. Arils color may be changed from typical cultivar color to light or white color).

The pomegranates must have reached an appropriate degree of development and ripeness in accordance with criteria proper to the varieties and to the area in which they are grown. The development and condition of the pomegranates must be such as to enable them:

- to withstand transport and handling; and
- to arrive in satisfactory condition at the place of destination.

According to Codex standards (CODEX STAN 310-2013) pomegranates are classified in three classes as defined below:

- "Extra" Class: Pomegranates in this class must be of superior quality. Since there are many cultivars, the fruits must carry the characteristic of the cultivar. They must be free of defects, with the exception of very slight superficial defects, provided these do not affect the general appearance of the produce, the quality, the keeping quality and presentation in the package. *Tolerance:* 5% by number or weight of fruits not satisfying the requirements of the class, but meeting those of Class I or, exceptionally, coming within the tolerances of that class.
- "Class I": Pomegranates in this class must be of good quality. Again, they must carry the characteristics of the cultivar. The following slight defects, however, may be allowed, provided these do not affect the general appearance of the produce, the quality, the keeping quality and presentation in the package:
  - slight defects in shape;
  - slight defects in coloring;

- slight skin defects including cracking;
- The defects must not, in any case, affect the arils of the fruit.
  *Tolerance:* 10% by number or weight of fruits not satisfying the requirements of the class, but meeting those of Class II or, exceptionally, coming within the tolerances of that class.

- "Class II": This class includes pomegranates which do not qualify for inclusion in the higher classes, but satisfy the minimum requirements specified above. The following defects, however, may be allowed, provided the pomegranates retain their essential characteristics as regards the quality, the keeping quality and presentation:
  - defects in shape;
  - defects in coloring;
  - skin defects including cracking;
  - The defects must not, in any case, affect the arils of the fruit.
    *Tolerance:* 10% by number or weight of fruits satisfying neither the requirements of the class nor the minimum requirements, with the exception of produce affected by rotting or any other deterioration rendering it unfit for consumption.

For all classes, 10% tolerance by number or weight of pomegranates corresponding to the size immediately above and/or below that indicated on the package is applied. Fruits may be sized by count, diameter or weight. The sizing method must be labelled accordingly on the package. As a guiding option for packaging, following table may be used:

**Table 16.** Packing sizes of pomegranate fruits

| Size Code | | Diameter (mm) | Weight (g) |
|---|---|---|---|
| 1 | A | >80 | >500 |
| 2 | B | 71-81 | 401-500 |
| 3 | C | 61-70 | 301-400 |
| 4 | D | 51-60 | 201-300 |
| 5 | E | 40-50 | 125-200 |

The categories in the above table are applied to all classes: "Extra", "Class I" and "Class II". The contents of each package must be uniform and contain only pomegranates of the same origin, variety, quality and size (if sized). Sales packages may contain mixtures of varieties of different colors and sizes provided they are uniform in quality and for each variety concerned, its origin. The visible part of the contents of the package must be representative of the entire contents. Pomegranate fruits must be packed in a way to protect the produce properly. The materials used inside the package must be new, clean, and of a quality such as to avoid causing any external or internal damage to the produce. "Extra" and "Class I" fruits must be packed in one-layer container/packages to avoid mechanical damage during storage or transport. The use of materials, particularly of paper or stamps bearing trade specifications is allowed, provided the printing or labelling has been done with non-toxic ink or glue.

- The containers/package must meet the quality, hygiene, ventilation and resistance characteristics to ensure suitable handling, shipping and preserving of the pomegranates. Packages must be free of all foreign matter and smell. Packages must be of a quality, strength and characteristic to protect the produce during transport and handling. Each container/package must be labelled to protect the right for consumers in the EU to access useful and appropriate information, Regulation (EU) No. 1169/2011 establishes the general principles, requirements, and responsibilities governing food information and in particular food labelling. Each package must bear the following information visible from the outside, or in the documents accompanying the shipment.
- Identification: Name and address of producer, packer and exporter. These informations may also be coded. For inspection purposes, the packer is the person or firm responsible for the packaging of the produce, not the employers. However, for inner control and traceability, the employer code may also be included to trace back if any problem arise due to packing.
- Name of the produce: Name of the cultivar.
- Origin of Produce: Country of origin and, optionally, district where grown or national, regional or local place name.
- Commercial Identification: Class, size (if sized), count, size or count range and optionally net weight.

However, packaging itself does not mean the products fulfill the requirements of buyers. There may be some other requirements asked by buyers. The requirements of buyers can be divided into three categories:

1. Musts: Requirements you must meet in order to enter the market, such as legal requirements; pesticide analysis, plant health, food control, marketing standard, contaminants, labelling and etc.;

2. Common requirements: These are those most of competitors have already implemented, in other words, the ones traders need to comply with in order to keep up with the market. GLOBALG.A.P. certificate, any other quality standards, food safety management system, etc.; and
3. Niche: Market requirements for specific segments, i.e., organic, fair trade, etc.

The EU has set maximum residue levels (MRLs) for pesticides in and on food products. EU MRL database can be used to find out the MRLs that are relevant for pomegranate. Products containing more pesticides than allowed will be withdrawn from the EU market. However, buyers in several countries may use MRLs that are stricter than the MRLs laid down in EU legislation.

Fruits and vegetables exported to the EU must comply with the EU legislation on plant health. Many countries in all over the world are applying similar standards for the pomegranates and for all other fruits and vegetables. All countries laid down phytosanitary requirements to prevent the introduction and spread of some organisms harmful to plants and plant products. For this reason, exporters must have a plant health certificate including the requirements of the imported country.

Nowadays, food safety is a top priority in all developed and developing countries. Thus, many countries are asking for food safety standards. GLOCALG.A.P. is the most commonly requested food safety certification in mainly European and all other countries. GLOBALG.A.P is a pre-farm-gate standard that covers the whole agricultural production process, from before the plant is in the ground to the non-processed product including harvesting, packing, labeling and exporting. Examples of other food safety management systems that can be required are BRC and IFS. However it must be kept in mind that different buyers may have different preferences for a certain food safety and management system. For this reason, it is important to check which is preferred in the target country before considering certification against one of these standards. It should be noted that quality refers to both food safety and food quality. However, in fact, food safety is a part of food quality. The standards that are most widely used by EU importers and traders are those developed by the United Nations Economic Commission for Europe (UNECE) and the Codex Alimentarius Commission (CAC).

## Packing

The pomegranates must be packed in a way that properly protects the product. The materials used inside the package must be clean and have high quality to not cause any damage on the fruit. Generally paper, stamps or plastic and carton violes are used. Individual stickers on the produce may be allowed if they neither leave visible traces of glue, nor lead to skin defects, when they are removed. Packages must be free of all foreign matter and odor. Packages/boxes may be made of wood, styro foam, cartoon and plastic, but corrugated cartoon box is the most popular rigid container. Stacking strength, length of storage, storage treatment and cost influence the choice of material. It is recommended to choice stronger materials against humidity.

The boxes for "Extra" and "Class I" fruits must be one-layer to protect minimum requirements for these categories during storage and transport. Packing pomegranates in more than one-layer boxes may cause damages at the point where pomegranates touch each other. On the other hand, violas or straw should be used at the bottom of the one-layer boxes to prevent fruits to touch each other.

Before or during sorting, the produce must be cleaned from the foreign and unwanted objects. In manual systems, workers do this job during sorting, but in automated systems, again a worker must stand before the line and control the fruits. This may be performed during final packaging by workers after sorting. However, it is better to protect the machines also from the foreign objects.

Pomegranate fruits are sensitive to some funguses, i.e. *Botryis cinera*. And, they must be treated with fungicides before packing. However, pomegranate fruits also sensitive against moisture inside the crown. For this reason, washing lines are not preferred for pomegranates due to the inefficiency of quick drying on line systems. Thus, once fruit is brought to packing house, it is recommended to wash the fruits with fungicide applied water and waited for one day for drying. Another way for crown drying is air pressuring into the crowns which also gives a control on the some ants and pests. After the harvest, fruits have to be sorted, packaged and transported. There are lots of technologies for the packing of pomegranates. New technology products offer users to sort fruits according to:

- *Color, sunburn and defect:* Sorting of fruits according to color is a very important step. Formerly, this practice was being carried by people by eye and hand. Nowadays, many companies offer solutions for this kind of sorting which reduces labor costs, eliminates mistakes and increase sorting speed. The only problem is the establishment costs of such kind of equipment. The fruits for color sorting are received by a fruit receiver and a high speed camera shoot fruit from various angles. The images from the camera are sent to the computer and computer determines the direction of the fruit according to given color parameters.

- *Diameter:* Grading of fruits into size groups is necessary for pomegranate fruits. Sorting products according to diameter is being done for many years. Size sorting is less precise than weight sorting but it is cheaper. Basically fruits pass through a belt type grader and roller and the fruits in adjusted dimensions are separated. The distance between the rollers is adjusted according to the preferred dimension. In this type of sorting, fruits fall down from the roller to another belt or packing line and skin and crown is damaged. Thus, this method is not preferred. Sorting according to diameter is being performed in some countries by eye and hand. New technology also offers some systems for non-destructive determination of horticultural produce dimensional size.
- *Shape:* As described before, shape disorder is common in pomegranates. Sorting of the fruits according to their shapes was done by workers manually, but new technology also offers users to adjust machines and machines can separate the fruits according to the wishes of user.
- *Weight:* Determination of fruit weight can be performed using a mechanical or an electronic weight sizer. Formerly, mechanical weight sizers were being used but nowadays, new and accurate machines are available for this job. These new machines can work on a line base and give quick results. The fruits which need to be sorted are transferred to the sorting line. Fruits there are received by a conveyer belt and the weight information of fruit is determined by a sensor. The analog weight information from sensor is transferred to digital information by switch. Thus, the computer receives the information and determines the direction of the fruit according to given parameters by the user.
- *Inner quality:* Determination of the inner quality of the pomegranates is the biggest issue for packers. As described within this book, Heart rot is a big problem for pomegranates where it has no obvious external symptoms. Professional workers may identify those fruits from outside by the abnormal skin color. The other way to identify this damage is the lesser weight of the fruits. Thus, if the weight sorting gives abnormal results, i.e., bigger fruits with less weight, it shows that the fruit is damaged by Heart rot.

Final packing of the produce into boxes must be done carefully to prevent any damages on the produce (Fig. 48). The dimensions of the boxes are generally 30 cm × 40 cm or 30 cm × 50 cm. The height of the boxes is changing depending on the sizes of pomegranates from 10 cm to 13 cm. For bigger sizes, higher boxes are being used. This is very important for the protection of the fruit quality. If the box height is not correct, bottom of the upper box would damage the fruits within the lower box. Boxes are then put on pallets, tied and sent to storage or transporting.

**Figure 48.** Single-line packing of pomegranate fruits into carton boxes

# CHAPTER 11

# POSTHARVEST BIOLOGY AND STORAGE

There are two important quality attributes for pomegranates, as in any other fruits, which are: external and internal quality attributes. External attributes include color, fruit size and shape whereas internal attributes include textural properties, total soluble solids, titratable acidity and flavor (brix/acid ratio), anthocyanin, vitamin C, sensory attributes, nutritional quality, functional quality and microbial quality. These attributes vary depending on cultivar, geographical conditions, climate and degree of maturity (Fawole et al. 2012a). Therefore, the most important aim of postharvest management is to determine suitable postharvest handling and storage practice to deliver the produce to the consumer in desirable condition. For this reason, first step in postharvest handling is the determination of the suitable handling and storage condition for the maintenance of the quality attributes.

## External Quality Attributes

### Color

Fruit skin color varies widely among cultivars from white to red. Darkly colored fruits tend to have the best flavor. Peel (and aril) color is the first impressions of consumers on the pomegranate fruit (Mena et al. 2011). The most popular cultivar in the world is the Wonderful cultivar which has red peel and aril color. Color measurements are generally performed with colorimeter (i.e., Minolta). Determination of skin and aril color is generally done according to the Hunter L*, a*, b* color scale. The Hunter L*, a*, b*, color scale is organized in a cube form. The L axis runs from top to bottom where maximum L is 100 which mean perfect reflecting diffuser, the white color. The minimum for L is zero which means black in color. The a* and b* axes have no specific limits. Positive a* is red and negative is green. On the other hand positive b* is yellow and negative b* is blue. Following is the diagram of the Hunter L*, a*, b*, color scale.

Al-Said et al. (2008) reported that L* value is changing from 55 to 87, where a* value chances from 15 to 35 and b* value change from 4 to 27. According to the researchers, the aril color values for same cultivars is

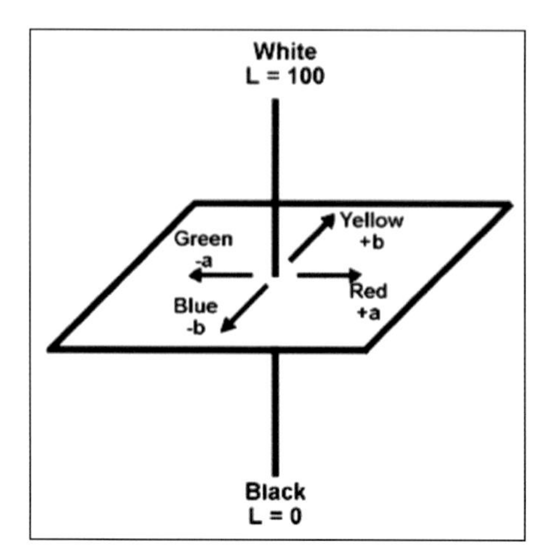


that; L value from 90 to 103, a* value from 1 to 3 and b* value from 0 to 3. Increase in the a* value makes the fruits more reddish. Thus, increase in the L value makes the fruits whiter. The highest L values from the study of Al-Said et al. (2008) were determined from the wild cultivars. Red color of the pomegranate fruit depends on anthocyanin concentration and on the chemical structure of the individual anthocyanin (Sepulveda et al. 2010).

Artés et al. (1998) reported that the L*, a*, and b* color parameters were higher in pomegranate fruit peel than in arils and juice. However, the authors observed no significant color difference in fruit peel and arils after 80 days of cold storage at 0°C and 5°C, respectively. However, Fawole and Opara (2013) reported significant decreases in the a* on Bhagwa cultivar when stored between 5°C and 10°C for up to 16 weeks of cold storage.

Derivatives of pelargonidin are responsible for red–orange colors where derivatives of delphinidin are responsible for blue and violet hues (Harborne 1982). Hernandez et al. (1999) reported that there are six anthocyanin pigments responsible for the red color of pomegranate juice. According to the researchers, those are: delphinidin 3- glucoside, 3,5-diglucoside, cyanidin 3-glucoside, 3,5-diglucoside, pelargonidin 3-glucoside and 3,5-diglucoside. They also reported that juice pigmentation increases during fruit ripening. In the early fruit-ripening stages, delphinidin 3,5-diglucoside is the main pigment, followed by cyanidin 3,5-diglucoside, while in the later stages, the monoglucoside derivatives cyaniding 3-glucoside and delphinidin 3-glucoside increases considerably.

**Fruit Size**

Fruit size can be from 5 cm on the small fruited cultivars to more than 18 cm on the larger fruited cultivars. According to Wetzstein et al. (2011)

small fruits have fewer than 300 arils per fruit; larger fruits have over 600 arils per fruit, and up to 985 in the largest fruits. After the characteristics of cultivar, flower quality is among the most important factors affecting fruit size in pomegranate. The production of large fruits requires flowers with adequate numbers of both functional ovules and a source of viable pollen. Pomegranate plant is characterized as having hermaphrodite and male flowers together, a condition called andromonecy (Holland et al. 2009). Ovule differentiation in pomegranates occurs before the opening of flower buds (Wetzstein and Ravid 2008). For this reason, adequate pollination is crucial for fruit size and development. According to Derin and Eti (2001), pomegranate flowers produce many pollens in anthers from 100 to more than 300 stamens per flower. Fruit size is also correlated with both number of arils and the pericarp. According to Wetzstein et al., (2011), percentage aril weight remained 50% of the total fruit weight regardless of fruit size. However, studies of Usanmaz et al. (2014) reported that aril weight as high as 68% (on Acco cultivar) of the total fruit weight. Aril weight of Wonderful cultivar, which is among the most important and commercial cultivars of the world is determined as around 60% by the same researchers. They also reported that the average fruit weight of Wonderful is 481.12 g, where the average weight of Acco cultivar is 350.31 g. However, it should be kept in mind that the geographical conditions, climate, pollination and orchard care are very important for the fruit size.

### Shape

Because of the globose or flattened shape of pomegranate fruits, the fruit diameter "D" (width) is always higher than the fruit length "L" (height). The fruit shape is determined as L/D. According to Usanmaz et al. (2014), the average fruit diameter of Wonderful is 107 mm where the length is 93 mm. The L/D of Wonderful cultivar is about 0.87. Radunic et al. (2015) conducted a study on 8 different cultivars and reported that the L/D ranges from 0.88 to 0.95 and reported that the lowest L/D score were obtained from wild cultivars. Fruit length (L), diameter (D), and volume of 'Crveni rani' and 'Dividiš' were reported to be highest, 0.95 and 0.91, respectively while the lowest values were found in wild pomegranate.

## Internal Quality Attributes

### Textural Properties

Textural properties of pomegranate fruit includes firmness, cohesiveness, gumminess, chewiness, springiness and resilience (Sayed-Ahmed 2014). Among these properties, firmness is the most known and useful properties which is measured with a penetrometer in Newton units. Fruit firmness is an important quality attribute for the postharvest quality of fruits and vegetables (Groos et al. 2002). Ismail et al. (2014) conducted a study on the textural properties of six pomegranate cultivars and reported that the peel firmness changes from 74.42 N to 79.98 N. Several studies reported that textural properties of pomegranate fruit change depending on storage conditions. According to Nanda et al. (2001), storage of Ganesh cultivar at 25°C, 15°C and 8°C resulted in decreases in fruit firmness after 1, 5 and 7 weeks, respectively. Mirdehghan et al. (2006a) reported that if Mollar de Elche stored at 2°C and 90% RH exhibits a significant decrease in firmness after 90 days. According to Mansouri et al. (2011), pomegranate fruits become less firm after 30 days of storage at 5°C. However, the decrease in the fruit firmness is also related with the moisture losses, which results hardening of the fruit peel. Tabatabaekoloor and Ebrahimpor (2013) conducted a study on the effects of storage conditions on the postharvest physic-mechanic properties of pomegranates and reported that the firmness is maintained significantly better when fruits were wrapped with polyethylene-film as compared to those wrapped with EPE-foam both at ambient and refrigerated conditions. Kazemi et al. (2013) reported a variation in the fruit firmness when they were treated with calcium chloride and sodium hypochlorite and stored at 5±1°C with 85±5% RH for 2 months. They reported that the firmness of treated fruits with sodium and calcium treatments was higher than the control fruits.

## Total Soluble Solids

Peel color of pomegranate fruit is not a reliable indicator for the ripening degree. For this reason, harvesting pomegranate fruits at correct time depends on some other factors, i.e., aril color, taste, and total soluble solids. According to some authors, it is crucial to combine some indices, i.e., peel color, aril pigmentation, total soluble solids and titratable acidity for the determination of fruit maturity (Cristosto et al. 2000, Martinez et al. 2006). However, in reality, total soluble solids content is the most used attribute for the determination of the fruit maturity. According to the maturity standards for Wonderful cultivar pomegranates grown in California and the total soluble solids content must be above 17%. They also reported that the titratable acidity (TA) level must be around 1.8%. However, these values are cultivar and geographical regional depended. Usanmaz et al. (2014) reported that Wonderful cultivar has average 21.45% total soluble content in Cyprus. According to same research, Acco has 17.29% and Herskovitz has 16.00% total soluble solids content averagely. Akbarpour et al. (2009) conducted a study on the total soluble solid contents of 12 cultivars in Iran and reported that the total soluble solids ranged from 15.17 to 22.03 (°Brix). On the other hand, Varasteh et al. (2009) evaluated five cultivars in Iran and observed total soluble solids varied from 16.60-18.26%. Total soluble solid contents of the fruits may be varied during storage, but it is also depending on the cultivar and storage conditions (Kader et al. 1984, Fawole and Opara 2012). Fawole and Opara (2013) reported significant decrease in total soluble solid contents with prolonged storage period for two cultivars: Bhagwa and Ruby, stored at 5°C, 7°C and 10°C at 92% RH for 12 weeks. Similar results have been reported by Artés et al. (1998) for Mollar de Elche cultivar stored at 0°C and 5°C at 95% RH for 80 days. It is easy to increase the number of publications reporting similar findings where Kader et al. (1984), reported significant decrease with increasing temperature and prolonged duration for Wonderful cultivar stored at 5°C for 16 weeks. The reason of the decrease in the total soluble content of the fruits is the degradation of sugars with prolonged storage period. But, if moisture loss is high during storage, it may lead to concentration of sugar inside the fruit and the total soluble content may increase (Köksal 1989). Findings of some studies support this phenomenon where Al-Mughrabi et al. (1995) observed an increase in total soluble content of 3 cultivars after 8 weeks of cold storage at 5°C, 10°C and 22°C. Similarly, Ghafir et al. (2010) reported that there was a significant increase in total soluble content of Shlefy cultivar when stored at 5°C and 7°C for 4 months.

**Titratable Acidity**

Titratable acidity (TA) of pomegranate juice differing among cultivars, growing region and maturity like total soluble solid contents (Fawole and Opara 2012). Kader et al. (1984) reported that pomegranate 'Wonderful' cultivar has an acidity content ranging between 1.11 to 1.58%. They also reported that storage conditions affect TA and cause a decrease at temperatures ranging between 0°C and 10°C for 16 weeks. However, on the other hand, Artés et al. (1998) reported no significant changes in TA for Mollar de Elche cultivar during storage at 5°C for 80 days. Moreover, he also reported that, 7 days of shelf-life after cold storage TA decreased significantly. Akbarpour et al. (2009) reported that titratable acidity of pomegranate juice is varying from 0.35% to 3.36%. Varasteh et al. (2009) evaluated five cultivars in Iran and reported a closer variation for the titratable acidity where he reported it as from 0.79 to 1.35%. Fawole and Opara (2013), reported a decreases in TA for the cultivars of Bhagwa and Ruby at 5°C, 7°C and 10°C for 4 months storage. On the contrary, Mirdehghan et al. (2006a) reported a significant increase in organic acids for Mollar de Elche stored at 2°C for 90 days. Similarly, Bayram et al. (2009) reported an increase in TA levels for untreated Hicaznar cultivar when stored at 6°C and 90% RH for 6 months.

### Brix/Acid Ratio

Brix/acid ratio is calculated by dividing the acid-corrected and temperature-corrected Brix by the percent titratable acidity w/w as citric acid (B/A ratio). This ratio is one of the most commonly used indicators of fruit maturity and juice quality indicators (Kimball 1991). Pomegranate flavor is generally known as sweet or sour depending on the cultivar. Therefore, the ratio between TSS and TA (TSS:TA or B/A) is the determination of this flavor. The level of sugar; fructose, glucose and sucrose determines the sweetness and organic acids; malic, tartaric and citric acids determines the sourness (Melgarejo et al. 2000). An accepted B/A ratio are not defined for the pomegranates but it is preferred to be above 13. The Brix/acid ratio is also called as maturity index (MI) by Hernandez et al. (1999) and is generally used to define the taste of fruit (Shwartz et al. 2009). Fawole and Opara (2013) reported that B/A ratio of Bhagwa cultivar is changing 16.68 at 54 days after full bloom to 39.19 at 140 days after full bloom and no significant changes until 165 days after full bloom (harvest). Similarly, Zarei et al. (2011) reported that B/A ratio of Rabbab-e-Fars cultivar vary from 3.73 to 14.48, respectively from 20 to 140 days after full bloom. Changes in B/A ratio are depending upon the changes in TA and TSS contents of fruit. According to Artés et al. (1998), there is no significant difference in juice TSS: TA ratio in Mollar de Elche cultivar stored at 5°C for 80 days. However they reported that after 7 days of shelf-life on 20°C, the ratio increased significantly. On the contracy, Fawole and Opara (2013) reported a decrease in TA and TSS during postharvest storage for Bhagwa and Ruby cultivars, and a significant increase in B/A ratio at most storage temperatures.

### Anthocyanin

There is a growing awareness on the health benefits of anthocyanin which causes an increase in the consumption of fruits including anthocyanin. According to Fischer et al. (2011), pomegranate is one of the major sources of anthocyanin. However, colorless polyphenols are predominant in pomegranates and anthocyanins has minor role in the biological activities of pomegranate fruits. Anthocyanins are natural pigments responsible for red, purple, and blue coloration in pomegranates (Gil et al. 1996, Artés et al. 1998). Types and content of anthocyanin vary among the pomegranate cultivars. The major anthocyanins reported in pomegranates were: delphinidin 3-glucoside, delphinidin 3,5-diglucoside, pelargonidin 3-glucoside, pelargonidin 3,5-diglucoside, cyanidin 3-glucoside, and cyanidin 3,5-diglucoside. Alighourchi et al. (2008) reported significant differences in anthocyanin levels among 15 Iranian pomegranate cultivars. They reported that some cultivars have very low anthocyanin contents, lower than 25.0 mg/l; however some cultivars, i.e., Mesri Torshe Kazeron and Torshe Mamoli Lasjer have anthocyanin contents more than 250.0 mg/l. They also reported that pasteurization of the pomegranate juice may cause the anthocyanin levels to decrease on some cultivars and increase on some other cultivars. Artés et al. (1998) reported that the total anthocyanin concentration in untreated fruits of Mollar de Elche cultivar increased during the 12 weeks storage duration at 0°C and 5°C in 95% RH. However, another research of the same author (Artés et al. 2000b) concluded with a contrary result where no change in anthocyanin concentration was observed for the same cultivar between harvest and shelf-life after 12 weeks. On the other hand, Varasteh et al. (2009) evaluated five cultivars in Iran and observed anthocyanin index which varied from 1.04-1.92%.

**Phenolic Concentration**

Phenolic compounds are responsible from many functional properties of pomegranate fruits (Gomez-Caravaca et al. 2013). Pomegranate fruit is known to have high antioxidant properties and phenolic contents (Madrigal-Carballo et al. 2009, Viuda-Martos et al. 2011, Zaouay et al. 2012). Fawole et al. (2012) reported that antioxidant activity of different pomegranate cultivars is also differing where in South Africa Bhagwa has highest antioxidant index of 95.7% and followed by Arakta (93.2%) and Ruby (79.9%). Li et al. (2006) reported that phenolic compounds which also accepted as antioxidant potential of fruits is higher in peel than in seeds and pulp. Similarly some studies reported that pomegranate peel has higher contents of phenolic compounds, such as: anthocyanins, ellagic acid glycosides, free ellagic acid, ellagitannins, punicalagin, punicalin and gallotannins (Saad et al. 2012). It is reported that phenolic compounds has a considerable decline during maturation (Kulkarni and Aradhya 2005). They reported that total phenolics has about 54.5% reduction during the initial stage of fruit development between 20 and 40 days after full bloom and this decrease continues until the 140th day (harvesting). Similarly, Weerakkody et al. (2010) reported a decline of about 50% for Wonderful cultivar. Another different study noted significant results where they reported that midwinter ripened fruit had higher concentration of total phenolics when compared with early summer, late summer and autumn ripened fruits (Borochov-Neori et al. 2011). As understood from the above mentioned studies, phenolic compounds decline during ripening. However, for human health, the higher the level of phenolic compounds means higher the total antioxidant activity of produce (Aviram et al. 2000, Gil et al. 2000, Tzulker et al. 2007). Mirdehghan et al. (2006b) reported that heat treatment can increase the phenolic compounds of pomegranate juice. According to their studies, contents of the phenolic compounds of heat treated fruits of Mollar de Elche cultivar was found to be 108.39 mg equivalent gallic acid 100 g-1 where the untreated pomegranate juice had 92.05 mg equivalent gallic acid 100 $g^{-1}$ when both stored at 2°C for 90 days. Fawole and Opara (2013), reported that total phenolic concentration of Bhagwa and Ruby cultivars showed a significant decline when stored at 5°C beyond 8 weeks. However, contrary results reported by Labbé et al. (2010) where significant increase noted on the phenolic contents of Chaca cultivar at 5°C for 12 weeks.

### Ascorbic Acid (Vitamin C)

Ascorbic acid is among the important components of pomegranate (Miguel et al. 2010). However it is believed that the storage duration of the fruits may cause a decline in the concentration of ascorbic acid. Furthermore some studies confirmed this idea. For example, Zarei et al. (2011) reported a significant decrease from 25.8 mg/100 g to 9.8 mg/100 g, from 20 day-old fruit to 140 day-old fruits, respectively. Thus, similar findings reported by Kulkarni and Aradya (2005) noted that the concentration of ascorbic acid declines rapidly in pomegranate juice (Ganesh cultivar) during initial stages of fruit maturity. On the contrary, some studies reported opposite findings, i.e., Shwards et al. (2009) where they reported a significant increase in the ascorbic acid concentration during the storage of Wonderful cultivar pomegranates.

### Volatile Composition

Volatile compounds of pomegranate fruit are thought to be low. Thus the odour and aroma of the fruit parts are low (Carbonell-Barrachina et al. 2012). Carbonell-Barrachina et al. (2012) reported 18 aromatic compounds in Spanish pomegranates whereas Melgarejo et al. (2011) identified 21 different aromatic compounds.

### Nutritional Quality

Not only are the arils of the fruits but the whole fruit is beneficial for human health. Pomegranate, as raw and juice, has a long history of nutritional value. However, the nutritional quality generally refers to the arils which are being directly eaten by human beings. Cultivar, fruit size, geographical region and production practices are directly related with the percent of arils and the pericarp. Percent of arils of the pomegranate fruits may be from 50% to 68% (Wetzstein et al. 2011, Usanmaz et al. 2014). However, lesser and higher percentages are also possible. As a general acceptance for the pomegranate arils; water constitutes about 85% and sugar 10%. Other 5% is mainly consisted from 1.5% pectin, organic acid such as ascorbic acid, citric acid, and malic acid, vitamins, polysaccharides, and important minerals compounds (Miguel et al. 2010). According to many studies, the highest amount of nutrient is found for potassium

**Table 17.** Nutrient contents of the edible portion and juice of Wonderful cultivar pomegranates (values per 100 g)

| Nutrients | Pom. raw | Pom. juice |
|---|---|---|
| ***Proximates*** | | |
| Water (g) | 77.93 | 85.95 |
| Energy (kcal) | 83 | 54 |
| Protein (g) | 1.67 | 0.15 |
| Total lipid (fat) | 1.17 | 0.29 |
| Carbohydrate, by difference (g) | 18.70 | 13.13 |
| Fiber, total dietary (g) | 4.0 | 0.10 |
| Sugar, total (g) | 13.67 | 12.65 |
| ***Minerals*** | | |
| Calcium, Ca (mg) | 10 | 11 |
| Iron, Fe (mg) | 0.30 | 0.10 |
| Magnesium, Mg (mg) | 12 | 7 |
| Phosphorus, P (mg) | 36 | 11 |
| Potassium, K (mg) | 236 | 214 |
| Sodium, Na (mg) | 3 | 9 |
| Zinc, Zn (mg) | 0.35 | 0.09 |
| ***Vitamins*** | | |
| Ascorbic acid: Vitamin C (mg) | 10.2 | 0.1 |
| Thiamin (mg) | 0.067 | 0.015 |
| Riboflavin (mg) | 0.053 | 0.015 |
| Niacin (mg) | 0.293 | 0.233 |
| Vitamin B-6 (mg) | 0.075 | 0.040 |
| Folate, DFE (µg) | 38 | 24 |
| Vitamin E (mg) | 0.60 | 0.38 |
| Vitamin K (µg) | 16.4 | 10.4 |
| ***Lipids*** | | |
| Fatty acids, total saturated (g) | 0.120 | 0.077 |
| Fatty acids, total monounsaturated (g) | 0.093 | 0.059 |
| Fatty acids, total polyunsaturated (g) | 0.079 | 0.050 |
| Cholesterol (g) | 0 | 0 |

(Al-Maiman and Ahmad 2002, Mirdehghan and Rahemi 2007, Gözlekci et al. 2011). Gözlekci et al. (2011) reported that the order of the mineral contents of the pomegranate fruit juice is as: K > P > Ca > Mg > Na > Mn > Zn > Fe > Cu. According to authors, the same order is found in fruit peel with the exceptions of Ca being higher than P. Fawole and Opara (2012) reported that general mineral composition (potassium, calcium, phosphorus, iron, magnesium, sodium, manganese, zinc, copper, nickel and selenium) of pomegranate arils (without seeds, only juice) varies from 0.14 to 6.9 ppm of fresh matter for 7 different South African cultivars. Pomegranate seed oils constitute about 12 to 20% of total aril weight. Pomegranate seeds are rich source of essential polyunsaturated fatty acids, i.e., linoleic and punicic acid (Ozgul-Yucel 2005). Seeds on the other hand contain protein, sugar, vitamins, fibers, minerals, pectin, isoflavones, polyphenols, phytoestrogen coumestrol, estrone and the sex steroid (Viuda-Martos et al. 2010). According to Grove and Grove (2008), 100 g pomegranate arils provide 72 kcal of energy, 1.0 g protein, 16.6 g carbohydrate, 1 mg sodium, 379 mg potassium, 13 mg calcium, 12 mg magnesium, 0.7 mg iron, 0.17 mg copper, 0.3 mg niacin and 7 mg vitamin C. Nutrient content of the edible portion and juice of Wonderful cultivar pomegranates are given in Table 17 (USDA 2015).

## Postharvest Deterioration

### Weight Loss

Weight loss is among the major postharvest problems of pomegranate fruits where it causes huge losses on weight and income for the marketers and cause hardening of the husk and browning of the rind in which they cause a reduction in the income for the marketers (Caleb et al. 2012). Apart from the visual appearance, considerable changes may occur on the textural quality of the fruits due to weight losses. According to Waskar (2011), shelf life of the pomegranate fruits at 21°C and 82% relative humidity is lower than 15 days. However, reduction in the storage temperature to around 5-7°C and increasing of the relative humidity over 90%, prolong the storage duration of the pomegranate fruits in terms of weight losses (Elyatem and Kader 1984, Köksal 1989, Küpper et al. 1995, Al-Mughrabi et al. 1995, Opara et al. 2008, Kazemi et al. 2013). Al-Mughrabi et al. (1995) reported that weight loss of pomegranate fruits may reach up to 32% if stored at 22°C for 8 weeks. On the other hand, Opara et al. (2008) noted that weight losses of Halow cultivar fruits may reach up to 16.42% at 21°C and 65% RH in 6 weeks, where the weight loss is only 3.85% if stored at 7°C and 95% RH for 6 weeks. Results of present studies show that the best way to reduce weight losses is decreasing storage temperature and increasing relative humidity. However, decreasing temperature causes chilling injury and increasing relative humidity provides a favorable condition for fruit pathogens, i.e., *Botrytis*.

### Chilling Injury

Chilling injury is another important postharvest storage problem which causes deterioration of pomegranates, after exposure to temperatures below 5°C for longer than 4 weeks. Thus, the incidence and severity of chilling injury depends on the storage duration and temperature. External symptoms of chilling injury include brown discoloration of the skin, necrotic pitting and increased susceptibility to decay. As an internal symptom; brown discoloration occurs on the white segments separating the arils and the color of the arils turns pale in color (Elyatem and Kader 1984). External and internal browning is related to oxidation of phenolics. Storaging fruits in an atmosphere with 2% oxygen at temperatures below 5°C may reduce chilling damage. According to Elyatem and Kader (1984), Wonderful cultivar pomegranate fruit is among the cultivars having high susceptibility to chilling injury if stored at temperatures below 5°C for more than 2 months. However, the severity or symptom of the chilling injury become more noticeable when transferred to ambient temperature after 2 months of cold storage. Artés et al. (2000a) reported that the best way to reduce chilling injury is the intermittent warming of fruits before storage. Moreover, Mirdehghan and Rahemi (2005) also reported that dipping fruits in 50°C water for 5 minutes significantly reduces chilling injury for Malas Yazdi and Malas Saveh cultivars when stored for 4.5 months at 1.5°C and 85±3% RH after dipping into hot water. Chilling tolerance of fruits stored at 2°C for 4 months may also be enhanced with the application of 2% calcium chloride and 2 mM spermidine to the fruit coats (Ramezanian and Rahemi, 2011).

### Husk Scald

Scald is a brown superficial discoloration restricted to the husk without observable internal changes on the arils. Husk scald increases susceptibility to decay and at advanced stages, the scalded areas became moldy. Late harvested fruits are known to be less susceptible than earlier harvested fruits (Ben-Arie and Or 1986). This physiological disorder is because of the oxidation of phenolic compounds on the husk when stored at temperatures above 5°C. Ben-Arie and Or (1986) noted a correlation between husk scald incidence and the amount of extractable o-dihydroxyphenols obtained from the affected husk. They also indicated that the most succesful control of husk scald in Wonderful cultivar pomegranates is the storage of fruits (late-harvested) in 2% oxygen at 2°C. However, the treatment resulted in accumulation of ethanol and caused off-flavors in the fruits. Nerya et al. (2006) reported that controlled atmosphere (CA) conditions of 2% $O_2$ and 0.6% $CO_2$ or 2% $O_2$ and 3% $CO_2$ and storage at 5°C reduced the incidence and severity of husk scald and internal chilling injury. Thus, they reported that the effects were enhanced at the higher $CO_2$ level. However, Diflippi et al. (2006) noted that pomegranates exhibited some scald after 4-6 months at 7°C and application of neither diphenylamine, at 1100 or 2200 $\mu L\ L^{-1}$, nor 1-methylcyclopropene at 1$\mu L\ L^{-1}$, alone or together reduced scald incidence and severity. In contrast, they reported that the three controlled atmosphere (CA) storage conditions tested (1 kPa $O_2$, 1 kPa $O_2$ + 15 kPa $CO_2$ and 5 kPa $O_2$ + 15 kPa $CO_2$) significantly reduced scald incidence and severity on pomegranates from both harvest dates for up to 6 months at 7°C.

## Decay

The major cause of decay on pomegranate fruit is gray mold which is caused by *Botrytis cinerea* (Fig. 49). There are some other fungi causing fruit rot including *Aspergillus niger*, *Alternaria* spp., *Penicillium* spp., *Coniella granati*, or *Pestalotiopsis versicolor* (Wilson and Ogawa 1979, Maclean et al. 2011). *B. cinera* infects the crown of young fruits in the field (sometimes the flower), host there and after harvest begin to spread from the crown to the rest of the fruit. On the other hand, *B. cinerea* is able to infect other stored pomegranates by spreading from infected fruit to adjacent healthy fruit (Caleb et al. 2012).

**Figure 49.** A view of fruit decay damage by *Botrytis cinerea*

The most favorable conditions for the development of gray mold are 5-10°C and relative humidity higher than 90%. The losses due to gray mold may reach up to 30% of harvested pomegranates without any control measure (Tedford et al. 2005). The application of fludioxonil (25% a.i. 60 g/100 l water) considerably reduces postharvest decay losses. However, to reduce the incidence of postharvest diseases should include a sanitizing chlorine wash prior to the application of fluodioxonil or any other fungicide (Palou et al. 2007). Spores of *Botrytis* hosts on previously infected fruits and trees in the orchard and on weeds. Thus they easily spread by wind. Spores landing on tissues germinate and produce an infection when there is free water on the plant surface. Decay develops quickly at the temperatures of between 18-24°C. Dust control and sanitation may help to reduce the postharvest incidence of disease.

## Respiration and Ethylene Production

Pomegranates are non-climacteric fruits which cannot continue the ripening process after harvesting (Kader et al. 1984). But, the pomegranate fruits are alive and continue respiratory process after harvest (Maguire et al. 2001). This process is essential to maintain biochemical, cellular organization and membrane integrity. Pomegranates are sensitive to variable temperatures which increase respiration and cause deterioration during postharvest handling. Elyatem and Kader (1984) reported trace amount (less than 0.2 μL/kg/h) of ethylene when fruits stored at 20°C for 2 weeks after storage in 0°C and 10°C for 3 months. They also reported a relatively low respiration rate (8 ml $CO_2$/kg/h). Opara *et al.* (2008) reported that the ethylene production (< 0.1μL/kg/h) of Helow cultivar pomegranate fruit may increase when stored at 21°C and 65% relative humidity for 6 weeks. They also reported that storing fruits at 7°C and 95% relative humidity significantly suppress the rate of ethylene production by over 63%. Elyatem and Kader (1984) indicated that Wonderful cultivar pomegranate fruit are not sensitive to ethylene exposure, where it may stimulate respiration at ≥ 1 μl/kg/h. However, Kader et al. (1984) reported that exposure of Wonderful cultivar pomegranate fruit to ethylene treatment at 20°C increases respiration rate, however cause no significant effects on fruit color, total soluble solids, or acidity. 10, 100 or 1000 ppm ethylene application to Wonderful cultivar for 2, 4 or 7 days at 20°C resulted with no significant effect on fruit external and internal attributes (Elyatem and Kader 1984). It is clear from the studies that pomegranate fruits should be picked when fully ripe to ensure the best quality for consumers.

## Storage Recommendation

Temperature and relative humidity are the two main elements of a successful storage of fruits and vegetables. However, as indicated above, there are 4 main causes of postharvest deterioration on pomegranates and these are differently influenced by temperature and relative humidity. Thus, it is very difficult to make a successful recommendation for the storage of pomegranate fruits. For example, temperatures below 5°C increases susceptibility to chilling injury but on the contrary temperatures above 5°C enhances husk scald and fruit decay and also increase weight losses. Moreover, relative humidity above 90% enhances fruit decay, while relative humidity below 90% increases weight loss and husk scald. According to the some studies (Kader et al. 1984, Al-Mughrabi et al. 1995, Küpper et al. 1995, Opara et al. 2008, Fawole and Opara 2013) most suitable temperature for the storage of pomegranate fruits is 5.5-6.5°C with a relative humidity of 85-95%. These storage conditions may help to store fruits up to 3 months or more depending on the cultivar. Application of fludioxonil is highly recommended for the control of fruit decay. Application of methyl jasmonate, calcium carbonate or some other chemicals has been tested for the prolongation of the storage duration by reducing the physiological deterioration on the fruits. However, none of them provide effective control. The newest and novel technology for the prolongation of the storage and shelf life of fruits is the modified atmosphere packing (MAP) technique. This treatment modifies the environmental conditions of fruit storage, effecting the fruit physiology and biochemistry and inhibiting the development of microorganisms contaminating the fruit surface, keeping the original physico-chemical quality of the fruit.

### Modified Atmosphere Storage

Modified atmosphere packaging is a technique where the surrounding atmosphere of the produce is removed entirely referred to as vacuum packaging, or the atmosphere is altered referred to as controlled atmosphere, or the surrounding atmosphere is modified continuously which refered to as modified atmosphere packing. In other words, if the adjustment of the atmospheric composition around the fruit is generated into gastight cold rooms or containers refered to as controlled atmosphere (CA) and if reached within hermetically sealed plastic packages referred to as modified atmosphere packaging (Artés and Hernandez 2006). In all cases, the goal is to extend the storage and shelf life of perishable produce by maintaining product quality. Modified atmosphere packing of fruits and vegetables is somehow differs from the fresh-cut produce and meats where horticultural crops are alive and continue to respire after harvest for vital biological reactions. When fresh fruits are harvested and packed they modify surrounding atmosphere by respiring $O_2$ and producing $CO_2$ (passive MAP). MAP slows the ongoing life processes by adjusting its surrounding environment. However, complete consumption of oxygen should be avoided to prevent anaerobic fermentation by selecting suitable packaging materials to permit $O_2$ to enter and $CO_2$ to leave the package. Another important concern for MAP technologies is the accumulation of ethylene in packs. Keeping ethylene concentration at low levels is mandatory in order to prolong shelf-life of ethylene sensitive produces. To rectify the deficiencies in passive MAP, certain additives may be incorporated into the packaging film to modify the headspace atmosphere, which is called as active Modified Atmosphere Packaging. By the activity of the absorbent materials, the headspace of the package is modified and it contributes to the extension of the storage and shelf-life of the produces.

To sum up; Modified atmosphere packaging (MAP) is a dynamic process of altering gaseous composition inside a package. The interaction between the respiration rate (RR) of the crop and the permeability of packing material are the two important elements of modified atmosphere packing, with no further control exerted over the initial gas composition (Farber et. Al. 2003, Mahajan et. al. 2007, Caleb et al. 2012). But, several factors directly influence the final package atmosphere including temperature, product weight and package surface area.

The most suitable conditions of pomegranates for succesfull storage vary depending on the cultivar (Kader et al. 1984, Köksal 1989, Artés et al. 1998). Artés et al. (1998) recommended a controlled atmosphere for Mollar de Elche cultivar at 5°C with 95% and with the concentration of 5% $O_2$ + 0% to 5% $CO_2$. In contrast, Kader (1995) recommended a gas composition of 3% to 5% $O_2$ + 5% to 10% $CO_2$ at 5°C. Artés et. al. (2000a) conducted a study on the modified atmosphere packing of Mollar de Elche cultivar stored at 2 or 5°C for 12 weeks. They used unperforated polypropylene (UPP) film of 25 μm thickness in modified atmosphere packaging (MAP) and perforated polypropylene (PPP) film of 20 μm thickness and conventional cold storage were applied as control treatments. They reported that all treatments suffered a decrease in total anthocyanins content at the end of shelf life; however MAP strongly reduced water loss and chilling injuries without incidence of decay. Artunes et al. (2007) reported that using modified atmosphere packing technique for the storage of pomegranate fruits, weight loss of fruits may be 7% when comparing the non-treated fruits with 13% when both stored at 5°C for 60 days. Bayram et. al. (2009) conducted a study to investigate the effects of streched film (12 μ) and MAP (8 μ) on the Hicaznar cultivar of pomegranate. They reported that fruits covered by streched film and MAP produced better storage conditions compared to uncovered fruits, and fruits stored for 2 months for streched film and 3 months for MAP without losing fruits quality. Kumar et al. (2012) investigated the effect of modified atmosphere packaging and polypropyle bags on the storage duration of Bhagwa cultivar at 4±1°C. They reported that shrinkage and decay was noticed in modified atmosphere package (Xtend®) bags after 90 days of cold storage.

# CHAPTER 12

# ARIL PRODUCTION

Consumption of pomegranate fruit is restricted due to the hassle of extracting the arils. Percentage of arils of the pomegranate fruits may be from 50% to 68% (Wetzstein et al. 2011, Usanmaz et al. 2014). For this reason, companies or enterpreneurs are extracting the arils by hand and supplying to consumers. Traditionally workers are peeling and extracting the arils from the pomegranate fruits which takes time and causes damages on the arils. Up to date, many studies have been conducted for the extraction of arils (Fig. 50) from the fruit. A prototype machine to separate arils and skin from fruits has been developed by MPKV, Rahuri. Also, a hand tool is designed by CIPHET, Ludhiana, India, for easy extraction of arils from fruits (Dhumal et al. 2014).

**Figure 50.** A view of the extracted arils from the fruit

First commercial machine was developed by Juran Metal Works Ltd., where they conducted research and development to build up technologies to extrat the arils from the fruits during 2000s. After receiving of the pomegranate fruits, fruits should be cured for a few days under shade to make the arils sensitive and thus can be easily separated from the peel. After that, fruits need to be washed to remove surface dirt. Thus, the whole fruit needs to be conveyed into machine where the skin is scored, the fruit is broken open and the arils are extracted with the aid of water and air. Extracted arils are immersed in cold water and washed. Air-blown dry fans are required to separate arils from all other fruit parts leaving them pristine and whole. Thus, the fruits are ready for packing and serving.

Packing of the pomegranate arils is the most important step to have longer shelf life by maintaining the quality of arils. The pomegranate arils are must be packed into containers with perforated plastic covering (special MAP package) that allows the arils to breathe while preventing oxygen from entering. Studies are being conducted for the prolongation of the storage duration of the packed ready-to-eat arils, but the duration is still limited to 14 days at 2-4°C. The other way of storing pomegranate arils is freezing them. By this way, arils may be stored for up to 1 year under –18°C. However, freezing arils is not an easy way in which it is therefore of paramount importance to freeze all arils individually and quickly to prevent cristalization and explosion of arils.

Several studies reported that modified atmosphere packing extend the shelf-life of minimally processed arils (Sepúlveda et al. 2000, López-Rubira et al. 2005, Ayhan and Eştürk 2009). Sepúlveda et al. (2000) studied the influence of different types of antioxidant solutions and packing materials as control on the quality of minimally processed pomegranate arils, cultivar Wonderful stored at 4±0.5°C for 14 days. They reported a slight browning observed on arils in all treatments, but highest in treatments without antioxidants. They noted that the use of semi-permeable films allowed successful storage for 14 days at 4±0.5°C, with good physical, chemical, and microbiological quality. García et al. (2000) studied the respiratory intensity of Mollar de Elche arils and the gas composition inside both a semi-permeable and an impermeable plastic at a storage temperature of 4°C for 10 days. They reported that the respiratory intensity is 30.8±0.4 ml $CO_2$/kg/h which is lower than the sliced oranges with 57.05±1 ml $CO_2$/kg/h from their study. They reported that the atmosphere within the semi-permeable plastic MAP packages was inadequate to prolong the shelf life of the minimally processed arils. In another study, López-Rubira et al. (2005) studied the effect of different UV-C radiation and passive MAP storage on shelf life, including chemical, sensory and microbial quality attributes of minimally processed arils of Mollar de Elche cultivar. They noted no significant differences between the control and UV-C treated arils and no observable interaction between the passive MAP and UV-C treatments. They suggested that the shelf life of fresh processed arils is at least 10 days Mollar de Elche cultivar pomegranate arils, harvested in early October, and stored at 1°C under MAP. Ergun and Ergun (2009) studied the effectiveness of 10 and 20% honey dip treatment on the quality and shelf life of minimally processed pomegranate arils of Hicaznar cultivar stored at 4°C in loosely closed plastic containers. They reported that honey treated arils had brilliant aroma throughout the 10 days storage period. Ayhan and Eştürk (2009) studied the effect of various gas compositions in active MAP of minimally processed pomegranate arils stored at 5°C. They reported that no significant change occurred in physicochemical attributes of arils during cold storage, while aerobic mesophilic bacteria were in the range of 2.30–4.51 log CFU $g^{-1}$. Caleb et al. (2013) studied the effects of storage temperature (5, 10 and 15°C) and passive modified atmosphere packaging (MAP) on the postharvest quality attributes, flavour attributes and microbiological quality of minimally processed pomegranate arils of Acco and Herskovitz cultivars. They reported that postharvest life of MA-packaged Acco and Herskovitz cultivar pomegranate arils was limited to 10 days due to fungal growth ≥2 log CFU $g^{-1}$ at 5°C. Hussein et al. (2015) investigated the effects of storage duration and number of perforations (P-0, -3, -6 and -9 per 160.1 $cm^2$) on the physicochemical quality attributes and microbiological quality of minimally fresh processed pomegranate arils of Acco Cultivar stored at 5°C for 15 days. They reported that highest $CO_2$ accumulation observed in non-perforated MAP, while $O_2$ concentration increased with increase in number of perforations in PM-MAP. For the results of total soluble solids, they reported that the highest decrease observed in P-9 PM-MAP arils from 15.4 to 13.1°Brix. They also indicated that highest counts of aerobic mesophilic bacteria (5.5 log CFU $g^{-1}$) and yeast and moulds (5.3 log CFU $g^{-1}$) determined from the P-0 and P- 9 PM-MAP. According to all results, they suggested that P-3 and P-6 PM-MAPs better maintained quality attributes of pomegranate arils than P-0 and P-9.

# CHAPTER 13

# JUICE PRODUCTION

Popularitiy of pomegranate juice is continuously increasing due to two reasons; 1) verification of the traditionally known health benefits of pomegranates, and 2) the hassle of extracting the arils for consumption. Aril percent of the pomegranate fruits may be from 50% to 68% (Wetzstein et al. 2011, Usanmaz et al. 2014) where the juice content of the arils may be up to 78% (Dhumal et al. 2014). The juice content of total fruit may be from 20.18% to 64.17% (Akbarpour et al. 2009, Martinez et al. 2006, Bartual et al. 2012, Pantelidis et al. 2012, Usanmaz et al. 2014) Environmental, postharvest, storage and processing factors also influence the juice quality and quantity. Juice is a more convenient way of consuming this highly valuable fruit. Juice yield of the different pomegranate cultivars also vary, with the varying color and antioxidant contents. The fruit disorders such as sun burn, husk scald and cracks on whole fruit reduces marketability and consumer acceptance. Processing of pomegranate allows using low quality fruits that cannot be commercialized, for the preparation of the new products. Despite of great demand and potential for pomegranate derived products, the industrial processing of pomegranate is new for the extraction of arils for industrial processing. Production of juice from the arils proved to be one of the important methods of value addition. The juice can be processed possible into the syrup, nectar, jelly and concentrate. Pomegranate juice can be used as an ingredient providing colour to the other products. Due to the increasing demand for pomegranate juice, processing of pomegranates to obtain juice by extracting the juice from arils or by crushing the intact fruit increased since 2000s. The main disadvantage of the crushing of intact fruit method is that the final product has astringent taste (presence of tannins) from the peel if no additional treatment is done to reduce astringency. This problem can be avoided by extracting the juice from the extracted arils. However, the antioxidant activity of the aril-pressed juice is lower due to the lower phenolic contents of the arils than the peel (Gil et al. 2000). After squeezing, numerous other optional steps then follow to strain, filter, purify and pasteurize the juice prior to bottling. Pomegranate juice may also be used to make concentrate but the public concern is growing about the "not from concentrate" juice. Concentration is done to about 70% TSS and stored frozen for subsequent thawing. To minimize the loss of taste and nutritional qualities, juice must be aseptically packed after concentration and stored in a deep frozen condition. Thus, the juice is diluted with pure water to about 17% TSS and marketed as concentrated juice. In other way, juice may be pasteurized to stabilize microbial activity. For clear juice, centrifuge and/or ultrafiltration process is necessary.

Given shelf lifes in the diagram (Fig. 51) are depending on the cultivar, geographical condition, preharvest applications, harvesting maturity, postharvest applications and the accuracy of the process. The hygiene of the production process is at utmost important for reaching maximum shelf life. Consumers generally prefer natural juice without centrifugal process, filtration or ultrafiltration. But, some consumers may ask for clear juices without pulps. In this case, centrifuge and/or ultrafiltration may be done. Centrifuge is an equipment which puts an object in rotation around a fixed axis working with sedimentation principle, and causing denser particles to move outward in the radial direction. Most of the pulps of the juice may be eliminated by centrifugal process. However, sometimes consumers ask

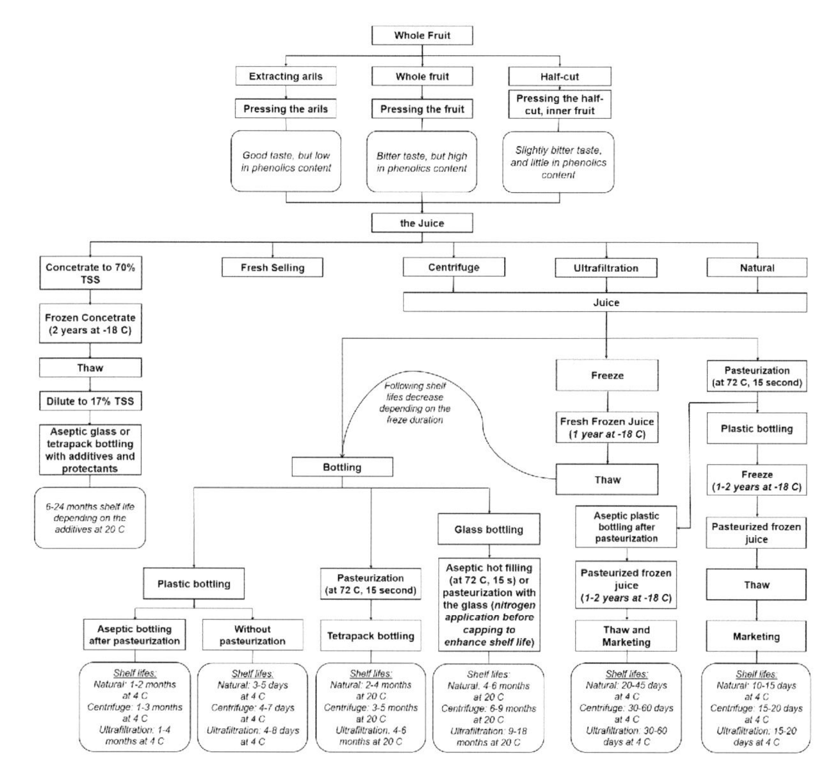


**Figure 51.** A diagram for the pomegranate juice processing phases

fully purified, clear juices. Ultrafiltration is a common process used for the production of clear fruit juice. Ultrafiltration is a variety of membrane filtration. Suspended solids and solutes of high molecular weight are retained at the filters while water and low molecular weight solutes pass through the membrane in the permeate. This method is commonly used for the clarification of the juice. However, it is highly important to rememder that all steps, like centrifugal process and altrafiltration causes a decrease and/or change in the phenolic compounds in juice, which are responsible from the health benefits, and color and total soluble solids (Valero et al. 2014).

Another important step for reaching to longer shelf lifes is the process of filling. In this case, aseptic filling is highly important. Aseptic filling is the process of filling the sterilized (pasteurized at 72 C for 15 second) juice into sterilized bottles. The process thus, has to be carried out under aseptic conditions. Pasteurization is another process which causes a decrease, but sometines increase in the quality attributes of pomegranate juice. It is a process where foods are heated to a specific temperature for a specific amount of time to kill some targeted harmful bacteria. Pasteurization is used in many products including milk, fruit juices, cider and beer. Pasteurization may cause low-level loss of some of the micronutrients, phenolic compounds and etc. Ascorbic acid (Vitamin C) is the most vulnerable vitamin to heat degradation during pasteurization. The pH of the product, level of phenolic compounds and pulp content are important for the determination of the right temperature and right holding time. The given temperature in the diagram is general and is not for all cases. According to FDA, 71.7°C for 15 second is needed for the 5-log reduction of oocysts of *Cryptosporidium parvum*, *Salmonela*, *Listeria monocytogenes* and *E. coli* in the fruit juice with a pH of lower than 4.0.

The foremost challenge in juice extraction is the extraction of pomegranate arils from the fruit which is time consuming by hand. The basic method for extraction of juice involves the cut opening of the fruit, seed separation and pressing. In another method, the fruits are quartered and crushed or the whole fruits may be pressed in hydraulic press and juice is strained out. A commercial method is now available by Juran Metal Works Ltd. and Bertuzzi to extract the arils from the fruit and press the arils to obtain high quality juice. On the other hand, pressing of the whole, half-cut or quarter-cut fruits is also available method for the extraction of juice. Spanos and Wrolstad (1992) reported that the pressing pressure should be less than 100 psi for the hydraulic extraction of juice to avoid undue yield of tannins from the rind. The phenolics contents of the pomegranate peel and juice too, are responsible for the color, astringency, cloudy appearance and bitterness of fruit.

Fining, ultrafiltration or clarification is one of the most important steps in fruit juice processing. However, with the increasing public awareness, there is a big challenge where many of the consumers began to demand un-clarified, pulpy juices. However, if whole fruit pressing system is used, high amount of tannin enters into the juice. The peel of the pomegranate fruit contains a very large amount of tannin and which may cause the juice to be undrinkable if the whole fruit is crushed or pressed with excessive pressure (Vardin 2000). Clarification helps to remove active haze precursors. In contrast, nutritionists recommend preserving these compounds during the fruit juice processing because of their health protective effects. However, clarification may be necessary to prevent the formation of cloudy appearance during storage and also to improve the taste of the product. Clarification reduces tannin contents which may cause bitter taste and may contribute to haze formation through the mechanisms involving condensation leading to the formation of the polymeric complexes and collected at the bottom of the fruit juice bottle when stored. Determination of the appropriate concentration of clarification agents is of utmost importance to achieve effective clarification. Pectinase enzyme plays important role in clarification of the fruit juices (Vilquez et al. 1981). Vardin (2000) reported the conventional heating of raw pomegranate juice, inactivates naturally present enzymes and destroys the vegetative microorganisms. He aslo noted that pasteurization may be applied after clarification. According to Apher et al. (2005), the most effective method for the removal of phenolic compounds from pomegranate juice is the conventional fining with gelatin (300 mg/l) and bentonite (300 mg/l) along with poly-vinyl-poly-pyrrolidone (PVPP). On the other hand, Vardin and Fenercioglu (2009) reported that the most effective method of clarification for the pomegranate juice is the application of 1 g/l gelatin before heat treatment. They noted that application of 1 g/l gelatin reduces the phenolic substances to an acceptable level, decreases turbidity, preserves anthocyanins and colour density. In contrast, Bayindirli et al. (1994) reported that 2 g/l gelatin is the most effective application for the reduction of tannin content in pomegranate juice. All these results suggest that the optimum amount of gelatin is varying depending on the cultivar. Ultrafiltration is another method for clarification of fruit juice which is a successful method for the removal of phenols. Neifar et al. (2009) reported that the laccase enzyme application to pomegranate juice cause three fold decrease in juice clarity with about 40% reduction of the total phenol content.

For the protection of the quality, color and nutrient composition of pomegranate juice, the selection of the packing material is very important. The main damages on the quality attributes are coming from light transmission and oxygen. Both have destructive effect on the anthocyanin

(Wasker and Deshmukh, 1995). Protection of the bottled juice from the direct sunligt is utmost important for the protection of the juice quality. Low density polyethylene (LDPE) and glass containers are commonly used materials for the packing of juices. LDPE is among the least harmful (may be harmless) plastic material used for the packing of fruit juice, water, milk and etc. This material can be washed, re-filled, heated and/ or freezed without causing any known damages on the stored juice. However, there is a growing concern on the use of plastic bottles and people prefer to consume glass bottled beverages. On the other hand, Sethi (1985) reported that glass containers are better when compared to high density polyethylene (HDPE) or polyvinylchloride (PVC) containers with regards to retention of anthocyanins, ascorbic acid. PVC is also known to include harmful contents for human health. The other important point is the storage temperature. For the retention of anthocyanins, reduction of enzymatic activity and slowing down the microbial activity in the pomegranate juice, storage temperature is recommended to be below 5°C. However, Ahire (2007) reported that the glass bottled pomegranate juice could be stored satisfactorily up to 3 months at 5±1°C conditions.

# CHAPTER 14

# POMEGRANATE AND HEALTH (REVIEW)

Pomegranate is an economically important plant (not only as a fruit) due to its health benefits and it is the major research subject of many studies. There are many studies conducted and revealed the health benefits of pomegranates. Pomegranate fruits are generally used as raw fruit, however it is also used as herbal healer since ancient times (Schubert et al. 1999, Sadeghi et al. 2009, Teixeira da Silva 2013). Singh et al. (1990) reported that pomegranate finds wide application in the traditional Asian medicines both in Ayurvedic and Unani systems. Production and consumption of pomegranate fruit is globally increasing due to the potential health benefits of the fruit, i.e., high antioxidant, anti-inflammatory effect and anti-carcinogenic (Gil et al. 1996, Malik et al. 2005, Jurenka, 2008). The pomegranate juice is one of the nature's most powerful antioxidants where it is reported to have 3 times higher antioxidant activity than red wine and green tea (Gil et al. 2000). Not only the juice, but almost all parts of the tree featured in medicine for thousands of years. According to Jurenka (2008), juice, seed oil, peel, leaves, flower, roots and bark of the pomegranate has different source of chemical constituents, which are beneficial for human health (Table 18).

**Table 18.** Constituents of different parts of pomegranates

| Plant part | Constituents |
|---|---|
| **The juice** | Anthocyanins, glucose, ascorbic acid, ellagic acid, gallic acid, catechin, minerals, aminoc acids, rutin, quecertin |
| **The seed oil** | Ellagic acid, sterols, 95% punicic acid |
| **The peel and pericarb** | Phenolic punicalagins, gallic acid, catechin, flavones, flavonones, anthocyanidins |
| **The leaves** | Tannins, flavone glycosides, luteolin, apigenin |
| **The flowers** | Gallic acid, urosolic acid, triterpenoids including maslinic and Asiatic acid |
| **The roots and bark** | Ellagitannins, piperidie alkaloids, punicalin and punicalagin |

According to Jurenka (2008), there are many therapeutic benefits of pomegranate which may be attributable to several mechanisms, but most researches have focused on its antioxidant, anticarcinogenic and anti-inflammatory properties. Followings are the review literature on the varying health benefits of pomegranates.

## Cardiovascular Health

Several studies conducted in vitro and animal & human trials carried to examine the effects of various pomegranate constituents on the prevention of low density lipoprotein (LDL) oxidation and cardiovascular disease (Aviram et al. 2000, Fuhrman et al. 2005, Sumner et al. 2005, Sezer et al. 2007, Basu and Penugonda 2009, Aviram and Resenblant 2013). Aviram et al. (2000) reported that pomegranate juice consumption decreased LDL susceptibility in humand to aggregation and retention and increased the activity of serum paraoxonase by 20%. In contrary, they reported that oxidation of LDL by peritoneal macrophages in mice reduced by up to 90% which is associated with reduced cellular lipid peroxidation and superoxide release. They noted that pomegranate juice with its antioxidative properties, have potential antiatherogenic effects in healthy humand and in atherosclerotic mice. Fuhrman et al. (2005) reported that pomegranate juice reduce celluar uptake of oxidized LDL and inhibit celluar cholesterol biosynthesis and has a direct effect on macrophage cholesterol metabolism. Sezer et al. (2007) compared the total phenol content and the antioxidant activity of pomegranate wine and red wine. They reported that the phenol levels of pomegranate wine were 4,850 mg/L gallic acid equivalents where the red wines' was 815 mg/L gallic acid equivalents. Thus the total antioxidant activity of pomegranate and red wine was reported to be 39.5% and 33.7%, respectively, by the authors. According to their studies, pure pomegranate wine demonstrated a greater antioxidant effect on diene level (110 +/– 4.6 micromol/mg of LDL protein) than pure red wine (124 +/– 3.2 micromol/mg of LDL protein). In conclusion, researchers suggest that pomegranate wine has potential protective effects toward LDL oxidation, and it may be a dietary choice for people who prefer fruit wines. Sumner et al., (2005) conducted a study to investigate the effects of pomegranate juice on patients who have ischemic coronary heart disease (CHD). 45 patients consumed 240 ml/ days pomegranate juice where same number of patients did not consumed for 3 months. The experimental and control groups showed similar levels of stress-induced ischemia. The extent of stress-induced ischemia after 3 months and decreased in the pomegranate group. This benefit was reported to be observed without changes in cardiac medications, blood sugar, weight or blood pressure in either group. Basu and Penugonda

(2009) reported that pomegranate juice is a polyphenol-rich fruit juice with high antioxidant capacity. According to researchers, pomegranate juice significantly reduces atherosclerotic lesion areas in immune-deficient mice and intima media thickness in cardiac patients on medications. Aviram and Resenblant (2013) reported that pomegranate juice substantially reduces macrophage cholesterol and oxidized lipids accumulation, and foam cell formation leading to attenuation of atherosclerosis development, and its consequent cardiovascular events. They also noted that pomegranate antioxidants are unique in their ability to increase the activity of the HDL-associated paraoxonase 1 (PON1), which breaks down harmful oxidized lipids in lipoproteins and in atherosclerotic plaques. To sum up, the unique characteristics of pomegranate antioxidants beneficially decrease blood pressure and help to protect cardiovascular disease. Daily consumption of one glass (250 ml) of pomegranate juice may be beneficial for human health.

## Hyperlipidemia

Lipids (fats in the blood) perform important functions in human body at proper levels. However, if excess they can cause health problems. The term hyperlipidemia means high lipid levels which may also be called as high cholesterol. Esmaillzadeh et al. (2006) conducted a pilot study with 22 type 2 diabetic patients (14 women and 8 men) and investigated the cholesterol-lowering effects of 40 g concentrated pomegranate juice for eight weeks. They reported statistically significant decreases in total cholesterol (from 202.4±27.7 mg/dL at control to 191.4±21 mg/dL at study conclusion), LDL cholesterol (124.4±31.9 mg/dL at control to 112.9±25.9 mg/dL at study conclusion), total/HDL cholesterol ratio (5.5±1.3 at control to 5.1±1.1 at study conclusion), and LDL/HDL ratio (3.4±1.2 at control to 3.0±0.9 at study conclusion). On the other hand, Huang et al. (2005) conducted a study with diabetic rats to explore the effects of PFLE on cardiac lipid metabolism in 13- to 15-weeks old Zucker diabetic rats. They reported that six weeds oral administration of pomegranate flower extract (500 ppm) reduced cardiac triglyceride content, accompanied by a decrease in plasma levels of triglyceride and total cholesterol, indicating improvement by pomegranate flower extract of abnormal cardiac TG accumulation and hyperlipidemia in this diabetic model.

## Hypertension

Hypertension (high blood pressure) is a chronic medical condition where the blood pressure in the arteries is elevated. Two measurements are used to express blood pressure, these are: the systolic and diastolic pressures. These are the maximum and minimum pressures, respectively, in the arterial system. When the left ventricle is most contracted, the systolic pressure occurs; and the diastolic pressure occurs when the left ventricle is most relaxed just before the next contraction. Normal blood pressure of adults at rest must be within the range of 100–140 mmHg systolic and 60–90 mmHg diastolic. Hypertension is present if the blood pressure is persistently at or above 140/90 mmHg for most adults (James et al. 2014).

Regular consumption of 100 cc pomegranate juice three times a week for one year is reported to improve systolic blood pressure, pulse pressure, triglycerides and HDL level. These beneficial outcomes were reported to be more pronounced among patients with hypertension, high level of triglycerides and low levels of HDL (Didi-Shema et al., 2014). Aviram and Dornfeld (2001) studied the effect of pomegranate juice consumption (50 ml, 1.5 mmol of total polyphenols per day, for 2 weeks) by hypertensive patients on their blood pressure and on serum angiotensin converting enzyme (ACE) activity. They noted 36% decrement in serum ACE activity and 5% reduction in systolic blood pressure.

## Anti-oxidant Properties

Antioxidants are molecules which are involved in the prevention of cellular damage or oxidation of other molecules, the common pathway for cancer, aging and a variety of diseases. There are number of free radicals in human body. Free radicals are highly reactive chemicals which have the potential to harm cells. They are formed naturally in the body and play an important role in many cellular reactions. They are created when an atom or a molecule either gains or loses an electron. Exposure to environmental toxins and ionizing radiation may cause the concentration of free radicals to abnormally increase. If the concentration of free radicals increases, they may be hazardous to the body and damage all major components of cells, including DNA, proteins and cell membranes. Thus, the damages on the body structure may play an important role in the development of cancer and other health conditions. Antioxidants are chemicals which have the ability to interact with the free radicals and neutralize them. Human body generally relies on external sources of antioxidants, primarily the diet, to obtain the required antioxidants against free radicals. Fruits, vegetables and grains are rich sources of dietary antioxidants. Some dietary antioxidants are also available as dietary supplements (Valko et al. 2007). Beta-carotene, lycopene and vitamins A, C and E (alpha-tocopherol) are some examples to the dietary antioxidants (Davis et al. 2012).

Due to the importance of antioxidants for human health, higher importance is given by researchers to determine the antioxidant activity of pomegranate components (Yasoubi et al. 2007, Ibrahium 2010, Gil et al. 2000, Elfalleh et al. 2012, Fawole et al. 2012b, Hassan et al. 2012).

Ibrahium (2010) carried out a research to investigate the antioxidant propertyies of pomegranate peel extract. He reported that the antioxidant activity of pomegranate peel extract at levels 400 and 800 ppm were 41.5 and 63.4%, respectively. Elfalleh et al. (2012) conducted a study to determine antioxidant contents of Gabsi cultivar. They reported that the antioxidant contents of the plant parts are as: peel > flower > leaf > seed. The antioxidant capacity value of peels was reported to be 7.50 ± 0.83 Trolox equivalent antioxidant capacities (TEAC) mg/g DW where antioxidant capacity of flowers was found to be 6.39 ± 0.83 TEAC mg/g DW. Less important values were obtained from leaves (4.16 ± 1.35 TEAC mg/g DW) and seeds (1.10 ± 0.23 TEAC mg/g DW). Yasoubi et al. (2007) and Fawole et al. (2012b) also suggested that pomegranate fruit peel could be accepted as a potential source of natural antioxidant agents as well as tyrosinase inhibitors. Rosenblat et al. (2006) noted that pomegranate extracts scavenge free radicals, and decrease oxidative stress and lipid peroxidation in animals. They studied with rats and mice and confirmed the antioxidant properties of a pomegranate extract made from whole fruit without the juice, showing 19% reduction in oxidative stress in mouse peritoneal macrophages, 42% decrease in cellular lipid peroxide content, and 53% increase in reduced glutathione levels. Karadeniz et al. (2005) conducted a comperative study to determine and compare the antioxidant activities of apple, quince, grape, pear and pomegranate. They assessed the total phenolic and flavonoid contents of those samples and reported that among the studied fruits, pomegranate had the highest (62.7%) antioxidant activity, followed by quince (60.4%), grape (26.6%), apple (25.7%) and pear (13.7%). On the other hand, Gil et al. (2000) reported that pomegranate juice has 3-fold higher antioxidant activity than that of red wine or green tea. Tzulker et al. (2007) studied the relationships between antioxidant activity, total polyphenol content, total anthocyanins content, and the levels of four major hydrolyzable tannins in four different juices/homogenates prepared from different sections of the fruit. They tested 29 different accessions for this purpose. The results showed that the antioxidant activity in aril juice correlated significantly to the total polyphenol and anthocyanin contents. Another study conducted by Hassan et al. (2012) suggested that the total antioxidant activity measured by FRAP assay of 32 pomegranate accessions growing in Egypt is ranging from 225.17 to 705.50 (mmol/100 g) and from 157.33 to 419.33 (mmol/100 ml) in peel and juice, respectively.

## Anti-carcinogenic Mechanisms

Cancer may be the biggest challenge for human health, can simply be called as a group of diseases involving abnormal cell growth with the potential to invade or spread to other parts of the body. In all types of cancer, some of the body's cells begin to divide without stopping and spread into surrounding tissues. Cancer may began almost anywhere in the human body. Normally, human cells grow and divide to form new cells when the body needs. Normally body cells grow old or become damaged, they die and new cells take their place. However, when cancer develops, this process breaks down and cells become more and more abnormal, where old or damaged cells continue to survive and new cells form without need. Thus, these extra cells can also divide without stopping and may form growths called tumors. Since cancer is among the most important diseases for human health, several studies conducted about the anti-carcinogenic activities of pomegranate and derived products (Lansky et al. 2005, Adams et al. 2006, Malik and Mukhtar 2006, Pantuck et al. 2006, Wang and Martin-Green 2014).

Lansky et al. (2005) investigated the effects of dissimilar biochemical fractions originating in anatomically discrete sections of the pomegranate fruit against proliferation, metastatic potential and phosholipase A2 (PLA2) expression of human prostate cancer cells in vitro. They reported that, various pomegranate extracts (juice, seed oil and peel) have a potential to inhibit protate cancer cell invasiveness and proliferation. They also reported that, combinations of pomegranate extracts from different parts of the fruit were more effective than any single extract. On the other hand, Adams et al. (2006) noted that phytochemicals of pomegranate fruit may inhibit cancer cell proliferation and apoptosis through the modulation of cellular transcription factors and signaling proteins. They reported that pomegranate juice (PJ) and its ellagitannins may inhibit proliferation and induce apoptosis in HT-29 colon cancer cells. Malik and Mukhtar (2006) studied the potential effects of dietary antioxidants of pomegranate fruit as candidate prostate cancer chemopreventive agents. They reported that pomegranate derived products possess strong antioxidant and anti-inflammatory properties. According to researchers, pomegranate fruit extracts may inhibit the growth of cell followed by apoptosis of highly aggressive human prostate carcinoma PC3 cells. They suggested that pomegranate consumption may retard prostate cancer progression, which may prolong the survival and quality of life of the patients. Similarly Pantuck et al. (2006) reported that pomegranate juice may have a promise as a therapy for prostate cancer, particularly recurrent type with rising PSA levels. They reported that LNCaP (prostate cancer cell) growth decreased about 12% in 84% of patients compared to control. Thus, an average 17.5% increase in apoptosis in 75% of patients was also reported. More recently, Wang and Martin-Green (2014) reported similar findings with the previous studies and indicated that pomegranate juice may significantly inhibit the growth of prostate cancer cells in culture. They also identified the most important components having role in the prevention of prostate cancer. According to researchers, these components are: luteolin, ellagic acid and punicic acid.

Kohno et al. (2004) investigated the effect of dietary pomegranate seed oil on the development of AOM-induced colonic malignancies and compared it with that of conjugated linoleic acid. They noted that administration of pomegranate seed oil in the diet significantly inhibited the incidence and the multiplicity of colonic adenocarcinomas, however a clear dose-response relationship was not observed by authors. Boateng et al. (2007) examined the effect of blueberries, blackberries, plums, mangoes, pomegranate juice, watermelon juice and cranberry juice on azoxymethane (AOM)-induced aberrant crypt foci (ACF) in Fisher 344 male rats. They reported that pomegranate fruit juice reduced the number of aberrant cryptic foci (ACF) of the colon by 91% in male rats. In another study, Kasimsetty et al. (2010) studied the colon cancer chemopreventive properties of pomegranate ellagitannins and their intestinal bacterial metabolites, urolithins, in HT-29 human colon cancer cells. They reported that the ellagitannins and urolithins released in the colon upon consumption of pomegranate juice in considerable amounts may potentially inhibit the risk of colon cancer development, by inhibiting cell proliferation and inducing apoptosis.

Along with the studies on the effects of pomegranates and/or derived-products on the prostate, colon and skin cancer, some studies also conducted on the breast cancer. Toi et al. (2003) indicated that pomegranate seed oil retard oxidation and prostaglandin synthesis and thus inhibit breast cancer cell proliferation and invasion, and promote spoptosis in breast cancer cells. Similar study by Mehta and Lansky (2004) reported that pomegranate fermented juice polyphenols caused a 42% reduction in the number of lesions compared with control. Moreover, according to researchers, purified chromatographic peak of fermented juice polyphenols and pomegranate seed oil each caused 87% reduction. Authors highlighted that purified compound peak and pomegranate seed oil has potential to prevent breast cancer. Clarifying the effects of pomegranate constituents on key hormones which are known to be involved in breast cancer may result in important information for consumers and light the way on the impact of diet on breast cancer risk. The proapoptotic effect of pomegranate extracts (40 µg/mL) was investigated by Jeune et al. (2005) on human breast cancer cells in combination with genistein. Authors suggested that the association of genistein and pomegranate might be more useful in association with anticancer drugs used for breast tumor. This is due to apoptosis induction and cell-growth inhibition of the combination was significantly higher than that of single compounds. Ellagic acid seems to exhibit apoptosis, inhibits activation of inflammatory pathways, and inhibits angiogenesis. Most of these assays conducted in animal models and need to be confirmed in humans (Sturgeon and Ronnenberg, 2010). Various parts of pomegranate tree have been stated to prove selective effects on lung, leukemia, stomach, bladder, oesaphagus and oral cancers (Jurenka, 2008; Apkıpar-Beyazıt, 2012; Husari et al., 2014).

## Anti-microbial Properties

The use of natural antimicrobials, chemical or synthetic agents is one of the oldest techniques for controlling microbial growth of bacteria and fungi (Mahmoudi et al., 2012). Therefore, essential oils could be an alternative to chemicals for control of postharvest phytopathogenic fungi on fruits or vegetables (Feng and Zheng, 2007). Several studies conducted about the antimicrobial activity of some common pomegranate cultivars (Voravuthikunchai et al., 2005, Reddy et al., 2007, Choi et al., 2009, Dahham et al., 2010, Sadik and Asker, 2014). For example; Voravuthikunchai et al. (2005) indicated that ethanol, chloroform and water extract of pomegranate showed high activity against strains of E. coli O157:H7. Similar findings reported by Reddy et al. (2007) where butanolic extracts were also tested against *E. coli*. They also reported anti-microbial effects against *Pseudomonas aeruginosa*, *Candida albicans* and *Cryptococcus neoformans*. Choi et al. (2009) investigated the in vitro and in vivo antimicrobial activity of pomegranate peel ethanol extract against 16 strains of Salmonella. The minimal inhibitory concentrations were in the range of 62.5 to 1000 $\mu g\ mL^{-1}$. According to researchers, pomegranate peel exthanol extract has the potential to provide an effective treatment for salmonellosis. Dahham et al. (2010) described the antibacterial and antifungal activities of pomegranate peel extract, seed extract, juice and whole fruit on the selected bacteria and fungi. The studies indicated that peel extract has shown highest antimicrobial activity compared to other extracts. Among the selected bacterial and fungal cultures, the highest antibacterial activity was recorded against *Staphylococcus aureus* and among fungi high activity against *Aspergillus niger* was recorded. Another study conducted by Sadik and Asker (2014) to examine the antioxidant activity of pomegranate peel and seed extracts. They used ethyl acetate, methanol, methanol, hexan, chloroform and water to extract the antioxidant rich fractions. According to authors, methanol extract of peels showed 83 and 81% antioxidant activity at 50 ppm using the DPPH model systems. Similarly, the methanol extract of seeds showed 22.6 and 23.2% antioxidant activity at 100 ppm using the DPPH model systems.

## Anti-inflammatory Activity

Inflammation is the body's attempt to self-protection by aiming to remove harmful stimuli, including damaged cells, irritants, or pathogens, and begin the healing process. If something irritating or harmful affects a part of human body, body gives a biological response and tries to remove it. The signs and symptoms of this biological response is inflammation, specifically acute inflammation, show that the body is trying to heal itself. Thus, inflammation is not infection, because infection is caused by a bacterium, virus or fungus, while inflammation is the body's response to it. Inflammation seems to be beneficial for human health, but it is out of control and it can damage the body. Plus, it's thought to play a role in obesity, heart disease, and cancer. Many studies have reported anti-inflammatory properties of pomegranate fruit (Jung et al., 2006, Lee et al., 2010, Sarker et al., 2012, Colombo et al., 2013). A whole pomegranate methanol extract was reported to inhibit the production and expression of TNFα in microglial cells, in which inflammation had been induced by lipopolysaccharide (Jung et al., 2006). Lee et al. (2010) analyzed 4 hydrolyzable tannins, punicalagin, punicalin, strictinin A, and granatin B, isolated from pomegranate by bioassay-guided fractionation. Each of them displayed a significant inhibitory effect on nitric oxide production in in vitro studies. Thus, researchers suggested that the components of pomegranate juice might appear to synergistically suppress inflammatory cytokine expression, where the fruits are widely used as an antipyretic analgesic in Chinese culture. Results of the Sarker et al. (2012) also signify the traditional uses of pomegranate for inflammation and pain. Different preparations of pomegranate, including extracts from peels, flowers, seeds, and juice, show a significant anti-inflammatory activity in the gut (Colombo et al., 2013).

## Anti-diabetic Properties

Diabetes is among the most common metabolic disease in the world and is still increasing. It simply describes a group of metabolic diseases in which the patient has high blood glucose (blood sugar), either because body do not produce insulin (Type 1) or the body does not produce enough insulin for proper function, or the cells in the body do not react to insulin (Type 2). Patients with high blood sugar will typically experience polyuria and they will become increasingly thirsty and hungry. One of the most important ways to control diabetes is through the diet, that's why pomegranate fruit and derivates play an aimportant role against diabetes. Numerous studies have described their anti-diabetic activity (Jafri et al., 2000, Das et al., 2001, Li et al., 2005, Esmaillzadeh et al., 2006, Das and Barman, 2012).

Jafri et al. (2000) reported that oral administration of an aqueous-ethanolic (50%, v/v) extract of male abortive flowers of pomegranate had a significant blood glucose lowering effect in normal, alloxan-induced diabetic rats and glucose-fed hyperglycemic. This effect of the extract was maximum at 400 mg/kg. Das et al. (2001) studied the hypoglycemic activity of pomegranate seed extract in rats made diabetic by streptozotocin. They reported that the seed extract (300 and 600 mg/kg, orally) caused a significant reduction of blood glucose levels in induced diabetic rats of 47% and 52%, respectively, after 12 hr. Another study suggested that pomegranate flower extract improves postprandial hyperglycemia in type 2 diabetes and obesity at Zucker diabetic fatty rats (Li et al., 2005). Another study in type 2 diabetic patients with hyperlipidemia noted that concentrated pomegranate juice (40 g daily for 8 wk) decreased cholesterol absorption, increased fecal excretion of cholesterol, had a beneficial effect on enzymes involved in cholesterol metabolism, significantly reduced total and LDL cholesterol, and improved total/HDL and LDL/HDL cholesterol ratios (Esmaillzadeh et al., 2006). On the contrary, Rashidi et al. (2013) reported that cholesterol and LDL-cholesterol concentrations decreased in patients consuming concentrated pomegranate juice (45 g/days for 3 months) when compated to control group but not significant ($p>0.05$). Das and Barman (2012) conducted a study to evaluate the anti-diabetic and anti-hyperlipidemic effects of ethanolic extract of leaves of pomegranates in alloxan-induced diabetic rats and they reported that pomegranates leaves possess significant antidiabetic and antihyperlipidemic activity.

## Anti-viral Properties

Four major polyphenols of pomegranate peel extracts; ellagic acid, caffeic acid, luteolin and punicalagin tested against influenza epidemics which cause numerous deaths and millions of hospitalizations each year. Results of that study identified punicalagin as the anti-influenza component, due to compound blocked replication of the virus RNA, inhibited agglutination of chicken RBC's by the virus, and had viricidal effects. Thus, researchers reported that pomegranate peel extracts inhibited the replication of human influenza A/Hong Kong (H3N2) in vitro (Haidari et al., 2009). On the other hand, Neurath et al. (2005) reported that pomegranate extract has microbiocidal effects on HIV-1. Howell and Souza (2013) noted that pomegranates have been known for hundreds of years for their multiple health benefits, including antimicrobial activity. According to authors, nearly every part of the pomegranate plant has been tested for antimicrobial activities, including the fruit juice, peel, arils, flowers, and bark and many studies have utilized pomegranate peel with success.

## Oral Health

Ployphenolic flavonoid content of pomegranate is thought to give a potential to pomegranates to be used as potential anti-plaque agent. It is also reported that pomegranate phytotherapy has potential to provide cost-effective and an indigenous solution in preventive dentistry (Ramesh and Shamin 2012). Herbal mouthwashes have been considered to be a more advantageous option to their chemical counterparts, for a long time, with the traditional use of: neem, honey bee extract and cranberry extract. Through a series of laboratory tests and animal trials Narayan et al. (2014) noted that pomegranate extract might reduce the clinical signs associated with chronic, inflammatory periodontitis, among other indications such as treatment of oral ulcers. It is also reported by DiSilvestro et al. (2009) that pomegranate contains agents, especially polyphenolic flavonoids and could be considered conducive to good oral health. Rinsing the mouth for 1 min with a mouth wash containing pomegranate extract effectively reduced the amount of microorganisms cultured from dental plaque. Vasconcelos et al. (2003) evaluated the use of a gel containing the extract of pomegranate as an antifungal agent against candidosis associated with denture stomatitis. They reported that a gel containing pomegranate extract applied 3 times per day for 15 days was effective for patients afflicted by candidiasis associated with denture stomatitis. A hydroalcoholic extract of pomegranate fruit (HAEP) was investigated by Menezes et al. (2006) for antibacterial effect on dental plaque microorganisms. Results indicated that hydroalcoholic extract was effective against *Staphylococcus, Streptococcus*, *Klebsiella*, and *Proteus species*, as well as *E. coli*. Kote et al. (2011) also indicated that pomegranate rinse is effective against dental plaque microorganisms. They reported significant reduction in the number of colony forming units of streptococci (23%) and lactobacilli (46%) by rinsing teeth with 30ml of pomegranate juice.

## Skin Health

Prolonged exposure to ultraviolet may cause serious adverse effects to human skin, i.e. premature skin aging, sunburn and skin cancer. Pomegranate derivatives show considerable protective effects on ultraviolet radiation. The major polyphenols in pomegranate, particularly catechin, play a significant role in its photoprotective effects on UVB-induced skin damage (Park et al. 2010). Aslam et al. (2006) reported that pomegranate peel (pericarp) is well-regarded for its astringent properties; the seeds for conferring invulnerability in combat and stimulating beauty and fertility. They suggested that aqueous extracts of pomegranate peel may be used for promoting regeneration of dermis, and pomegranate seed oil for the promoting regeneration of epidermis.

Hora et al. (2003) investigated the possible efficacy of pomegranate seed oil on the skin cancer of mice. They used 5% pomegranate seed oil as a protectant on the tested mice and reported that 20 wk of promotion affected the average number of skin tumors (16.3 per mouse) when compared with control (20.8 per mouse). They also reported that pomegranate seed oil caused 17% eduction in ODV activity and authors highlighted the potential of pomegranate seed oil as a safe and effective chemopreventive agent against skin cancer. Syed et al. (2006) suggested that pomegranate fruit extract treatment of normal human epidermal keratinocytes (NHEK) inhibits UVB-mediated activation of MAPK and NF-KB pathways. They reported that extracts were prepared from the edible part of fruit with acetone, and treatments showed to inhibit UVA-induced phosphorylation of STAT3, ERK1/2 and AKT1 in human epidermal cells. In another study by Afaq et al. (2009) it is reported that pretreatment of human reconstituted skin (EpiDermTM FT-200) with pomegranate-derived products inhibited UVB-induced CPDs and 8-OHdG as well as protein oxidation.

## Obesity

Obesity is defined as abnormal or excessive fat accumulation on human beings which presents a risk to human health. Body mass index (BMI) is the simple measure of obesity. It is calculated when a person's weight (in kilograms) divided by the square of her or his height (in metres). A person with a BMI of ≥30 is generally considered obese. A person with a BMI ≥25 is considered overweight. Overweight and obesity are major risk factors for a number of chronic diseases, including diabetes, cardiovascular diseases and cancer. According to WHO, at least 2.8 million people die each year as a result of being overweight or obese in the world. The worldwide prevalence of obesity is continuously increasing throughout the world. In 2008, 10% of men and 14% of women in the world were obese, compared with 5% for men and 8% for women in 1980. An estimated 205 million men and 297 million women over the age of 20 were obese – a total of more than half a billion adults worldwide in 2008 (WHO 2015).

Cerda et al. (2003) studied the influence of pomegranate extract in female rats following exposure to repeated oral administration of a 6% punicalagin-containing diet for 37 days. A significant decrease in feed consumption and body weight of the animals during the early part of the study was noted.They reported that no significant differences were found in treated rats in any blood parameter analyzed with the exception of urea and triglycerides. Although the reason for the decrease is unclear, it could be due to the lower nutritional value of the punicalagin-enriched diet with respect to the standard rat food. Histopathological analysis of liver and kidney corroborated the absence of toxicity. Thus, researchers indicated that inspite of the lack of toxic effect of punicalagin in rats during the 37 days period, it must be studied for humans with a lower dose and during longer period of intake due to the high punicalagin content of pomegranate-derived foodstuffs. In a similar study, Lei et al. (2007) investigated the antiobesity effects of pomegranate leaf extract in a mouse model of high-fat diet-induced obesity. They reported that the extract inhibited the development of obesity and hyperlipidemia. When the weight of the high-fat diet group was 20% higher than the normal diet group, the animals were treated with 400 or 800 mg/kg/day of pomegranate leaf extract for 5 wk. The treated groups showed a significant decrease in body weight, energy intake and various adipose pad weight percents and serum, TC, TG, glucose levels and TC/HDL-C ratio after 5 wk treatment. Thus authors noted that pomegranate leaf extract may be a novel appetite suppressant that only affects obesity owing to a high-fat diet. Adnyana et al. (2014) conducted another similar study to evaluate in vivo and in vitro assay from the pomegranate leaves ethanol extract as the anti-obesity. Results showed that pomegranate leaves ethanol extract at a dose 50 mg/kg and 100 mg/kg showed a significant decrease of body weight, faeces index, total fat index, food index, and lee's index compared to control mice. Authors reported that the pomegranate leaves ethanol extract may be a potentially therapeutic alternative in the treatment of obesity caused by a high-fat diet. In a review study by Al-Muammar and Khan (2012), it is reported that studies in the field of obesity are still limited and need more attention that would help in understanding the preventive and protective roles pomegranate extracts have on obesity

## Erectile Dysfunction

Erectile dysfunction refers to not getting and/or maintaining an erection. In some cases the penis becomes partly erect but not hard enough to have sex properly. It may also be called as impotence. The pomegranate fruit juice is also known to be very helpful in treating issues of erectile dysfunctions (Bhowmik et al. 2013). For the determination of the potential effects of pomegranate juice on the erectile dysfunction in male, a study conducted by Forest et al. (2007). They conducted a randomized, double-blind trial with 53 men having mild-to-moderate impotence. Authors reported that blindly consumption of pomegranate juice, or placebo, for 4 weeks. caused men more likely to have improved scores. They reported that of the 42 subjects who demonstrated improvement in GAQ scores after beverage consumption, 25 reported improvement after drinking pomegranate juice. On the other hand, Azadzoi et al. (2005) noted that longterm pomegranate juice intake (3.87 ml) in a rabbit model, increased intracavernous blood flow and improved erectile response and smooth muscle relaxation in erectile dysfunction. Thus, they reported that intake of pomegranate juice prevented erectile tissue fibrosis. According to Zhang et al. (2010) most cases of erectile dysfunction are associated with oxidative stress risk factors They conducted a study with the mice by assigning animals into two groups receiving pomegranate extract antioxidants in drinking water or tap water as placebo. They continued tests for 8 wk and after that they noted that pomegranate extract significantly improved erectile activity and smooth muscle relaxation of the atherosclerotic group in comparison with the atherosclerotic group receiving placebo. But authors reported that they did not normalize to the age-matched control levels.

## Sperm Quality

Pomegranate juice consumption is thought to increase in epididymal sperm concentration, sperm motility, spermatogenic cell density and also decrease the abnormal sperm rate (Türk et al., 2008). Dkhil et al. (2013) reported that both methanolic extract of pomegranate peels and pomegranate juice have potent antioxidant activity by reducing lipid peroxidation and nitric oxide formation in testis tissues of rats. They also reported that methanolic extract of pomegranate peels and pomegranate juice caused high elevation in male sex hormones as testosterone, follicular stimulating hormone and luteinizing hormone. Another similar study by Mansour et al. (2013) aslo reported that administration of pomegranate extract could modify sperm characteristics and antioxidant activity of adult male wistar rats. Pomegranate juice also reported to be a protectant of testicular toxicity by increasing blood testosterone and improveming sperm quality (Luangpirom et al., 2013). Similarly Zeweil et al. (2013) reported that diets containing 1.5, 3.0 and 4.5% of pomegranate peel extract improved sperm motility by 28, 34 and 49%, increased sperm total output by 37, 69 and 102% and reduced dead sperm by 24, 32 and 64% compared to the heat stressed control rabbit.

## Alzheimer's Disease

Alzheimer's is a type of dementia which causes problems with memory, thinking and behavior. Dementia word describes a set of symptoms that can include memory loss and difficulties with thinking, problem-solving or language. Symptoms usually develop slowly and get worse over time, becoming severe enough to interfere with daily tasks. Although there are no proven ways to delay onset or slow progression of Alzheimer's disease, studies suggest that diet can affect risk. The neuroprotective properties of pomegranate polyphenols were evaluated in an animal model. Transgenic mice (APP (sw)/Tg2576) received either pomegranate juice or sugar water control from 6 to 12.5 months of age. It is reported that mice treated with pomegranate juice learned water maze tasks more quickly and swam faster than controls. Mice treated with pomegranate juice had significantly less (approximately 50%) accumulation of soluble Abeta42 and amyloid deposition in the hippocampus as compared to control mice. These results suggest that further studies to validate and determine the mechanism of these effects, as well as whether substances in pomegranate juice may be useful in Alzheimer's disease (Hartman et al. 2006). Thus, Dr Olumayokun Olajide reported that regular intake and regular consumption of pomegranate has many health benefits including prevention of neuro-inflammation related to dementia which slows down the progression of the disease (Weller, 2014)

## Potential Drug Interactions

Due to the increase in the studies about the health benefits of pomegranates, the consumption of this magic fruit is continuously increasing throughout the world and is commonly used in folk medicine for a wide variety of therapeutic purposes. The fruit is a rich source of several ellagic acid, pectin, tannins, flavonoids, anthocyanins, etc. Nagata et al. (2007) noted that pomegranate juice ingestion inhibits the intestinal metabolism of tolbutamide without inhibiting the hepatic metabolism in rats. They also reported that pomegranate juice inhibited human CYP2C9 activity and furthermore increased tolbutamide bioavailability in rats. Saruwatari et al. (2008) reported that pomegranate juice showed potential to inhibit the sulfo-conjugation of 1-naphthol in Caco-2 cells. It has been suggested that some constituents of pomegranate juice, most probably punicalagin, may impair the metabolic functions of the intestine (specifically sulfoconjugation) and therefore might have effects upon the bioavailability of drugs. Adukondalu et al. (2010) investigated the effect of pomegranate juice pre-treatment on the transport of carbamazepine across the rat intestine. The control and pomegranate juice (10 ml $kg^{-1}$ for 7 g) pre-treated rats were sacrificed and isolated the intestine. Authors reported that there was a significant difference in the transport of carbamazepine from the intestinal sacs of pretreated with pomegranate juice and control and it seems that pomegranate juice might have induced CYP3A4 enzymes and hence drug is extensively metabolized. Çelik et al. (2014) noted that there would be an interaction of pomegranate juice with warfarin. They reported that a 54-year-old male patient who recently started warfarin treatment 1×5 mg/day presented to the emergency department for INR control on the sixth day of treatment. Pomegranate juice and warfarin were interrupted and an INR fall was observed. Thus, authors reported that when taking the medical history of patients with warfarin use, all drug and food consumption that may affect the metabolism of warfarin should be investigated.

# CHAPTER 15

# POMEGRANATE TRADE

Pomegranate is one of the world's oldest known ancient fruits and it is also a mysterious fruit with mythical associations. But, the fruit and juice have gained commercial significance since 2000s with the verification of its health effects by the scientific works in all over the world. The fruit, flowers, bark and leaves contain bioactive photochemical which are antimicrobial, reduce blood pressure and act against serious diseases such as diabetes and cancer (Gil et al., 2000; Lansky et al., 2005; Jurenka, 2008; Turk et al., 2008; Haidari et al., 2009). Nowadays there is an increasing worldwide demand for this fruit owing to its superior pharmacological and therapeutic properties. The most popular variety in all over the world is Wonderful and is followed by Mollar de Elche, Hicaznar and Bhagwa. With the increasing popularitiy of the pomegranates, production areas have been increasing in all over the world. Unfortunately, there is no worldwide information about pomegranate production or sales. Food and Agriculture Organization of the United Nations (FAO) did not include pomegranates into its database where the United States Department of Agriculture (USDA) stopped collecting pomegranate data in 1989. European Union was presenting data about the trade of pomegranates until 2013, but they also stopped data presentation. According to USAID Report (2008) total production was around 2.5 million tonnes in 2008. According to same data and the author's knowledge, India (900.000 tonnes) is the world's largest producer of pomegranates, followed by Iran (800.000 tonnes). Other countries like Turkey (376.000 tonnes in 2014 by Turkish Statistical Institute 2015), USA (110.000 tonnes), Spain, Iraq, Afghanistan, Azerbaijan, Uzbekistan, Israel, Tunisia, Morocco, China, Cyprus, Egypt, Peru, Chile, Argentina and South Africa also produce pomegranate. In Europe, the main production area is in the eastern part of Spain, where the season is from October to January (with the help of cold storage). It is calculated that there are 13.000 ha of farmland dedicated to pomegranates in Spain, producing a total of 36 thousand tons a year (CBI 2015). Similarly, no data is available for the total real consumption of pomegranate for the Eropean Union. The most clear information is from CBI (2015) noted that 67.000 tonnes of pomegranates (imports minus exports) were added to the apparent consumption of pomegranate in the EU in 2013. Not only the pomegranates, but the pomegranate-derived products are gaining in popularity. Soft drinks (juices and fruit drinks) and food supplements dominate new innovative products with pomegranate as a main ingredient or pomegranate flavor launched in the supermarkets in.

Pomegranates can be consumed as fresh fruit or used in fruit juices, syrup, molasses or wine spirits (Fig. 52). The fruits may also be used as a decoration agent and as cattle feed due to their richness in fibre and antioxidants.

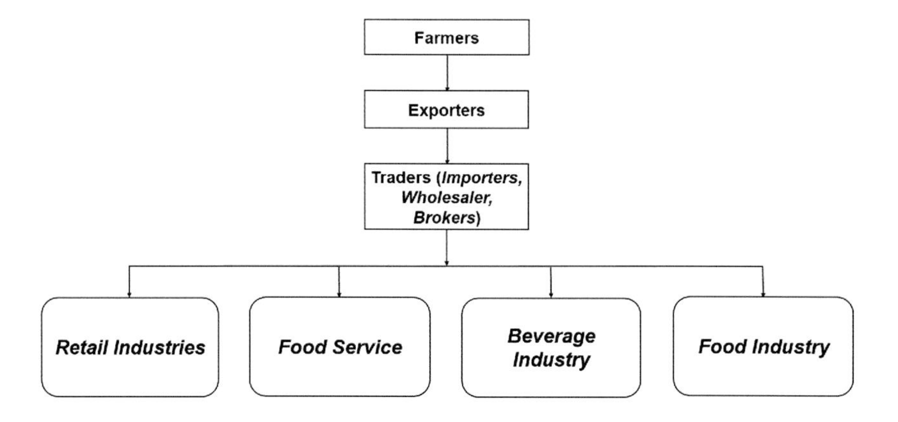


**Figure 52.** A diagram for the pomegranate handover from farmers to consumers

Due to the scientific confirmation of the health benefits, technology development of aril extraction machines, availability of ready-to-eat fresh arils and natural juice packs, pomegranates are expected to be among the first 10 fruits of consumption in coming years. Availability of pomegranate fruit in the global market is covering all year round with the different geographical advantages of the countries in the world and the availability of new technologies for postharvest storage (Table 19).

The European countries imports pomegranates year-around. The highest import volumes from outside the EU are in May (4.54 €/kg). The average monthly prices for pomegranates vary between 1.40 €/kg and 4.54 €/kg. The yearly average price for Holland, which is the main place of EU's import and export and has data for whole year, was 2.69 €/kg.

Table 20. shows monthly wholesale prices for some selected European countries for the year of 2014 (ITC, 2015). For the previous years, Rymon (2012) reported that average yearly pomegranate prices in Europe fluctuated between 2.50 €/kg and 3.50 €/kg from 2002 to 2011. The prices which indicated by Rymon are average yearly prices.

**Table 19.** Global availability of pomegranates

| Country | Jan | Feb | Mar | Apr | May | Jun | Jul | Aug | Sep | Oct | Nov | Dec |
|---|---|---|---|---|---|---|---|---|---|---|---|---|
| India | | | | | | | | | | | | |
| Iran | | | | | | | | | | | | |
| Turkey | | | | | | | | | | | | |
| USA | | | | | | | | | | | | |
| Spain | | | | | | | | | | | | |
| Israel | | | | | | | | | | | | |
| Argentina | | | | | | | | | | | | |
| Peru | | | | | | | | | | | | |
| Chile | | | | | | | | | | | | |
| S. Africa | | | | | | | | | | | | |

**Table 20.** Average monthly wholesale prices (€/kg) in some selected European countries

| Countries | Jan | Feb | Mar | Apr | May | Jun | Jul | Aug | Sep | Oct | Nov | Dec |
|---|---|---|---|---|---|---|---|---|---|---|---|---|
| **Belgium** | 2.00$^{\pm}$ | 1.88$^{\pm}$ | 1.88$^{\pm}$ | *N/A* | *N/A* | *N/A* | *N/A* | *N/A* | 2.08$^{\pm}$ | *N/A* | 1.88$^{\pm}$ | 1.50$^{\pm}$ |
| **France** | *N/A* | *N/A* | 2.25$^{\pm}$ | 5.00* | *N/A* | 2.25$^{\infty}$ | *N/A* | *N/A* | *N/A* | *N/A* | 2.40$^{\pm}$ | 2.50$^{\pm}$ |
| **Holland** | 1.92$^{\alpha}$ | 1.92$^{\alpha}$ | 4.00* | 4.29* | 3.86* | 2.63* | 2.37* | 3.18$^{\pm}$ | 2.00$^{\beta}$ | 2.14$^{\pm}$ | 2.20$^{\pm}$ | 1.81$^{\pm}$ |
| **Sweden** | 1.44″ | 1.47$^{\beta}$ | 1.63″ | 2.10″ | 4.54$^{\infty}$ | 4.44* | *N/A* | *N/A* | 3.64* | 3.68$^{\pm}$ | 1.46″ | *N/A* |
| **UK** | 1.89″ | 1.56″ | *N/A* | *N/A* | 3.74* | 3.45$^{¢}$ | 1.89$^{\beta}$ | 2.37$^{\beta}$ | 3.38$^{¢}$ | 2.14$^{\beta}$ | 1.41$^{\beta}$ | 1.64″ |

$^{\pm}$from Israel, $^{\alpha}$from Iran, ″from Turkey, $^{\beta}$from Egypt, *from Peru by air, $^{\infty}$from South Africa, $^{¢}$from Chile

On the other hand, consumer prices are differing from the wholesale prices where it may reach up to 1.2-3.0 fold of it. However, of course lower priced pomegranates are available in the markets. The prices, received by farmers, are on the other hand is about the one third of the wholesale price. However this ratio is depend upon the cultivar and quality of the product, packing technique used by exporters, preferences and buyig power of the target audience and way of transport. There are significant differences in monthly prices of fresh pomegranates during the main harvest season of the northern hemisphere (September to February, with the help of cold storage) and the main supply season of the southern hemisphere (March to July).

It could be concluded from the trade data that the demand for the pomegranates is continuously increasing with the slight increase or decrease in price. However, it is also clear that the production of this highly valuable fruit is increasing rapidly than the consumption. Thus, people who produce and/or market pomegranate should pay more attention on the awareness of human beings about the health benefits of pomegranates to increase consumption. Nowadays, pomegranate is classified in the European media as a "superfruit" with its higher levels of antioxidants and nutritional contents. On the other hand, health benefits of the pomegranates are main drivers for market success where European consumers embrace healthy and tasty fruits. The main goal of the farmers and/or exporters must be to reaching target markets during the time of off-season. For this reason, the using of new technologies and natural extracts for the prolongation of the storage duration of fresh fruits is utmost important. The other important way is to production of natural pomegranate juice with the high technologies. However, there are of course some important issues need to be considered and meet before entering into the market. These are:

- **Certification and other legal requirements**. Certification applies to Good Agricultural Practices (GAP) and legal requirements mainly apply to MRLs, food safety and plant health issues.
- **Nonlegal (buyer) requirements**. There may some requirements of buyers, about environmental and worker safety, packing materials, and etc.
- **Sweet taste and appearence.** Generally reddish fruits with big size and sweet taste are preferred. However, it may vary depending on the target audience.
- **Convenience**. An average time for preparing a meal or a snack has become shorter in developed countries. For this reason, fruits with easy-peeling characteristics are becoming more popular. Pomegranate fruit is not considered to be a fruit that is ready to be eaten quickly; however extraction of arils of some cultivars, i.e., Wonderful is easy.

On the other hand, extraction of arils with high tech equipments and packing in MAP packages is possible.

The supply quality and quantity, availability of pomegranates in the market and way of transport are the key factors affecting fruit prices. The term "quality" here includes the correct color, correct size, correct shape, free from pests, disease and physiological disorders and correct postharvest quality attributes (weight losses, chilling injury, husk scald and decay). Therefore, the product "quality" at the time of marketing, is utmost important for the success. Therefore, the main important issue for the farmers, marketers and/or scientists is the prolongation of the storage duration of pomegranates by maintaining the fruit "quality" to reaching target markets during the time of off-season. Alteration and/or adjustment of the temperature and relative humidity is very important for the storage of fruits. However, due to the characteristics of weight loss, chilling injury, husk scald and fruit decay, it is very difficult to determine a suitable temperature and relative humidity. In this case, modified atmosphere packing of pomegranates is very important for the prolongation of the storage duration. Combination of modified atmosphere packing with the application of fludioxonil seems to be the best way for the longer storage of pomegranates by maintaining fruit quality. However, nowadays not only the demand is increasing for the pomegranates but public awareness on the health is increasing. In this case, many scientists are working on the effects of natural extracts and/or plant and animal derived products on the postharvest quality of pomegranates. Several studies reported considerable effects of propolis and researchers noted positive effects on the prolongation of storage duration and maintaining quality of star ruby, navel orange, papaya and dragon fruit (Özdemir et al., 2010; El-badawy et al., 2012; Zahid et al., 2013; Ali et al., 2014). Eukaliptus (Tzortzakis, 2007), Aloe vera (Marpudi et al., 2011; Nabigol and Asghari, 2013) and thyme oil (Abdolahi et al., 2010; Fatemi et al., 2012; Baiea and El-Badawy, 2013) are also known to have positive effects on the storage duration of fruits and vegetables. No studies have been conducted on the effects of those or similar products for the pomegranates up to date. It is highly recommended to study the effects of those and other similar natural products on the storage duration and postharvest quality of pomegranate fruits, alone or in combination with modified atmosphere packing, the other environmental and safety way of storing. Success for the prolongation of the storage duration of pomegranates by using natural products, would give a power to farmers and/or exporters to increase their profitability, which is the main goal of farming.

Final recommendation for the readers:

"Market the product before you produce"

# BIBLIOGRAPHY

Adams, F. 1849. *Genuine works of Hippocrates*. William Wood and Co., New York.

Adams, L.S., N.P. Seeram, B.B. Aggarwal, Y. Takada, D. Sand, and D. Heber. 2006. Pomegranate juice, total pomegranate ellagitannins, and punicalagin suppress inflammatory cell signaling in colon cancer cells. *Journal of Agricultural Food Chemistry* 54: 980-985.

Adnyana, I.K., E.Y. Sukandar, A. Yuniarto, and S. Finna. 2014. Anti-obesity effect of the pomegranate leaves ethanol extract (Punica granatum L.) In high-fat diet induced mice. *International Journal of Pharmacy and Pharmaceutical Science* 6(4): 626-63.

Adukondalu, D., Y.S. Kumar, Y.V. Vishnu, R.S. Kumar, and Y.M. Rao. 2010. Effect of pomegranate juice pre-treatment on the transport of carbamazepine across rat intestine. *DARU-Journal of Faculty of Pharmacy* 18(4): 254-259.

Afaq, F., M.A. Zaid, N. Khan, M. Dreher, and H. Mukhtar. 2009. Protective effect of pomegranate-derived products on UVB-mediated damage in human reconstituted skin. *Experimental Dermatology* 18(6): 553-561 [Abstract only].

Ahire, D.B. 2007. Studies on extraction, packaging and storage of pomegranate (Punica granatum L.) juice cv. Mridula. MSc Thesis submitted to Mahatma Phule Krishi Vidyapeeth, Rahuri (Maharashtra), India.

Ahmed, N.H. and S.M. Abd-Rabou. 2010. Host plants, geographical distribution, natural enemies and biological studies of the citrus citrus mealybug, *Planococcus citri* (Risso) (Hemiptera: Pseudococcidae). *Egyptian Journal of Biological Sciences* 3(1): 39-47.

Akbarpour, V., K. Hemmati, and M. Sharifani, 2009. Physical and chemical properties of pomegranate (Punica granatum L.) fruit in maturation stage. *American-Eurasian Journal of Agricultural and Environmental Science* 6(4): 411-416.

Ali, A., C.K. Cheong, and N. Zahid. 2014. Composite Effect of Propolis and Gum Arabic to Control Postharvest Anthracnose and Maintain Quality of Papaya during Storage. *International Journal of Agricultural Biology* 16: 1117-1122.

Alighourchi, H., M. Barzegar, and S. Abbasi. 2008. Anthocyanins characterization of 15 Iranian pomegranate (Punica granatum L.) varieties and their variation after cold storage and pasteurization. *European Food Research and Technology* 227: 881-887.

Al-Maiman, S.A. and D. Ahmad. 2002. Changes in physical and chemical properties during pomegranate (Punica granatum L.) fruit maturation. *Food Chemistry* 76: 437-441.

Al-Muammar, M.N. and F. Khan. 2012. Obesity: The preventive role of the pomegranate (Punica granatum). *Nutrition* 28: 595-604.

Al-Mughrabi, M.A., M.A. Bacha, and A.O. Abdelrahman. 1995. Effects of storage temperature and duration on fruit quality of three pomegranate cultivars. *Journal of King Saudi University* 7: 239-248.

Al-Said, F.A., L.U. Opara, and R.A. Al-Yahyai. 2008. Physico-chemical and textural quality attributes of pomegranate cultivars (*Punica granatum* L.) grown in the Sultanate of Oman. *Journal of Food Engineering* 90: 129-134.

Artés F., P. Gómez, and F. Artés-Hernández. 2006. Modified atmosphere packaging of fruits and vegetables. *Stewart Postharvest Review* 5(3): 1-13.

Artés, F., F.A.R. Villaescusa, and J.A. Tudela. 2000a. Modified Atmosphere Packaging of Pomegranate. *Food Chemistry and Toxicology* 65(7): 1112-1127.

Artés, F., J.A. Tudela, and M. Gil. 1998. Improving the keeping quality of pomegranate fruit by intermittent warming. *European Food Research and Technology* 207: 316-321.

Artés, F., J.A. Tudela, and R. Villaescusa. 2000b. Thermal postharvest treatment for improving pomegranate quality and shelf-life. *Postharvest Biology and Technology* 18: 245-251.

Aslam, M.N., E.P. Lansky, and J. Varani. 2006. Pomegranate as a cosmeceutical source: Pomegranate fractions promote proliferation and procollagen synthesis and inhibit matrix metalloproteinase-1 production in human skin cells. *Journal of Ethnopharmacol* 103: 311-318.

Aviram, M. and L. Dornfeld. 2001. Pomegranate juice consumption inhibits serum angiotensin converting enzyme activity and reduces systolic blood pressure. *Atherosclerosis* 158: 195-198. [Abstract only]

Aviram, M. and M.S. Rosenblat. 2013. Pomegranate for Your Cardiovascular Health. *Clinical Implications of Basic Research* 4(2): 1-12.

Aviram, M., L. Dornfeld, M. Rosenblat, N. Volkova, M. Kaplan, R. Coleman, T. Hayek, D. Presser and B. Fuhrman. 2000. Pomegranate juice consumption reduces oxidative stress, atherogenic modifications to LDL, and platelet aggregation: Studies in humans and in atherosclerotic apolipoprotein E-deficient mice. *The American Journal of Clinical Nutrition* 71: 1062-1076. [Abstract only]

Ayhan, Z. and O. Eştürk. 2009. Overall quality and shelf life of minimally processed and modified atmosphere packaged "ready-to-eat" pomegranate arils. *Journal of Food Science* 74: 399-405.

Azadzoi, K.M, R.N. Schulman, M. Aviram, and M.B. Siroky. 2005. Oxidative stress in arteriogenic erectile dysfunction: prophylactic role of antioxidants. *Journal of Urology* 174(1): 386-393.

Baiea, M.H.M. and H.E.M. El-Badawy. 2013. Response of Washington navel orange to thyme and clove oils as natural postharvest treatments under cold storage conditions. *Journal of Applied Sciences Research* 9(7): 4335-4344.

Barkai-Golan, R. 2001. *Postharvest Diseases of Fruits and Vegetables: Development and Control*. Elsevier, New York.

Bartual, J., H. Valdes, J. Andreu, A. Lozoya, J. Garcia, and M.L. Badenes, 2012. Pomegranate improvement through clonal selection and hybridization in Elche. *Options Mediterraneenes* 103: 71-74.

Basu, A. and K. Penugonda. 2009. Pomegranate juice: a heart-healthy fruit juice. *Nutritional Review* 67(1): 49-56.
Bayindirli, L., S. Sahin, and N. Artik, 1994. The effects of clarification methods on pomegranate juice quality. *Fruit Processing* 9: 267-270.
Bayram, E., O. Dundar, and O. Ozkaya. 2009. Effect of different packaging types on storage of 'Hicaznar' pomegranate fruits. *Acta Horticulturae* 818: 319-322.
Ben-Arie, R. and E. Or. 1986. The development and control or husk scald on 'Wonderful' pomegranate fruit during storage. *Journal Horticultural of Science* 111: 395-399.
Bhowmik, D., H. Gopinath, B.P. Kumar, S. Duraivel, G. Aravind and K.P.S. Kumar. 2013. Medicinal Uses of Punica granatum and Its Health Benefits. *Journal of Pharmacognosy and Phytochemistry* 1(5): 28-35.
Blumenfeld, A., F. Shaya, and R. Hillel. 2000. Cultivation of pomegranate. *Options Mediterraneenes* 42: 143-147.
Boateng, J., M. Verghese, L. Shackleford, L.T. Walker, J. Khatiwada, S. Ogutu, D.S. Williams, J. Jones, M. Guyton, D. Asiamah, F. Henderson, L. Grant, M. DeBruce, A. Johnson, S. Washington and C.B. Chawan. 2007. Selected Fruits Reduce Azoxymethane (AOM)-induced Aberrant Crypt foci (ACF) in Fisher 344 Male Rats. *Food and Chemical Toxicology* 45(5): 725-732 [Abstract only].
Borochov-Neori, H., S. Judeinstein, M. Harari, I. Bar-Ya'akov, B.S. Patil, S. Lurie and D. Holland. 2011. Climate effects on anthocyanin accumulation and composition in the pomegranate (Punica granatum L.) fruit arils. *Journal of Agricultural and Food Chemistry* 59: 5325-5334.
Caleb, O.J., U.L. Opara, and C.R. Witthuhn. 2012. Modified atmosphere packaging of pomegranate fruit and arils. *Food and Bioprocess Technology* 5: 15-30.
Caleb, O.J., U.L. Opara, P.V. Mahajan, M. Manley, L. Mokwena, and A.G.J. Tredoux. 2013. Effect of modified atmosphere packaging and storage temperature on volatile composition and postharvest life of minimally-processed pomegranate arils (cvs. 'Acco' and 'Herskawitz'). *Postharvest Biology and Technology* 79: 54-61.
Carbonell-Barrachina, A.A., A. Calín-Sánchez, B. Bagatar, F. Hernández, P. Legua, R. Martínez-Font and P. Melgarejo. 2012. Potential of Spanish sour-sweet pomegranates (cultivar C25) for the juice industry. *International Journal of Food Science and Technology* 18: 129-138.
CBI, 2015. CBI Product Fact Sheet: Fresh Pomegranates in the European Market 'Practical market insight into your product'. CBI (Centre for the Promotion of Imports from developing countries) 17p. http://www.cbi.eu/sites/default/files/study/product-factsheet-pomegranates-europe-fresh-fruit-vegetables-2014.pdf (access on April 23, 2015)
Çelik, G.K., G.P. Günaydın, N.Ö. Doğan, M.M. Dellül, and H.Ş. Kavaklı. 2014. Pomegranate Juice and Warfarin Interaction: A Case Report (Nar Suyu ve Varfarin Etkileşimi: Olgu Sunumu). *Journal of Academic Emergency Medicine Case Reports* 5: 66-68.
Cerda, B., J.J. Ceron, F.A. Tomas-Barberan and J.C. Espin. 2003. Repeated oral administration of high doses of pomegranate ellagitannin punicalagin to rats for 37 days is not toxic. *Journal of Agricultural and Food Chemistry* 51: 3493-3501. [Abstract only]

Choi, J.G., O.H. Kang, Y.S. Lee, H.S. Chae, Y.C. Oh, O.O. Brice, M.S. Kim, D.H. Sohn, H.S. Kim, H. Park, D.W. Shin, J.R. Rho, and D.Y. Kwon. 2009. In vitro and in vivo antibacterial activity of Punica granatum peels ethanol extract against salmonella. *Evidence-Based Complementary and Alternative Medicine* 17: 1-8.

Colombo, E., E. Sangiovanni, and M. Dell'Agli. 2013. A Review on the Anti-Inflammatory Activity of Pomegranate in the Gastrointestinal Tract. *Evidence-Based Complementary and Alternative Medicine* Article ID 247145, 11p.

Cristosto, C.H., E.J. Mitcham, and A.A. Kader. 2000. Pomegranate: recommendations for maintaining postharvest quality. *Produce Facts Postharvest Research and Information Centre*, University of California, Davis, USA. http://postharvest.ucdavis.edu/PFfruits/ Pomegranate/ (access on January 29, 2015)

Crites, A.M. 2004. Growing Pomegranates in Southern Nevada. *University of Neveda, Cooperative Extension* 4p. http://www.agmrc.org/media/cms/fs0476_DF280E5 E65077.pdf (access on April 29, 2015)

Das, A.K., S.C. Mandal, S.K. Banerjee, S. Sinha, B.P. Saha, and M. Pal. 2001. Studies on the hypoglycaemic activity of Punica granatum seed in streptozotocin induced diabetic rats. *Phytotherapy Research* 15(7): 628-629 [Abstract only]

Das, S., and S. Barman. 2012. Antidiabetic and antihyperlipidemic effects of ethanolic extract of leaves of Punica granatum in alloxan-induced non-insulin-dependent diabetes mellitus albino rats. *Indian Journal of Pharmacology* 44: 219-224.

Davis, C.D., P.A. Tsuji, and J.A. Milner. 2012. Selenoproteins and Cancer Prevention. *Annual Review of Nutrition* 32: 73-95 [Abstract only].

Derin, K. and S. Eti. 2001. Determination of pollen quality, quantity and effect of cross pollination on the fruit set and quality in the pomegranate. *Turkish Journal of Agriculture and Forestry* 25: 169-173.

Dhumal, S.S., A.R. Karale, S.B. Jadhay and V.P. Kad. 2014. Recent Advances and the Developments in the Pomegranate Processing and Utilization: A Review. *Journal of Agriculture and Crop Science* 1(1): 01-17.

Didi-Shema, L., B. Kristal, S. Sela, R. Geron, and L. Ore. 2014. Does Pomegranate intake attenuate cardiovascular risk factors in hemodialysis patients? *Nutrition Journal* 13-18.

DiSilvestro, R.A. and D.J. DiSilvestro. 2009. Pomegranate extract mouth rinsing effects on saliva measures relevant to gingivitis risk. *Phytotherapy Research* 23: 1123-1127 [Abstract only].

Dkhil, M.A., S. Al-Quraishy, and A.E.A. Moneim. 2013. Effect of Pomegranate (Punica granatum L.) Juice and Methanolic Peel Extract on Testis of Male Rats. *Pakistan Journal of Zoology* 45(5): 1343-1349.

El-Badawy, H.E., M.H. Baiea, and A.A. Eman. 2012. Efficacy of propolis and wax coatings in improving fruit quality of Washington navel orange under cold storage. *Research Journal of Agriculture and Biological Sciences* 8: 420-428.

Elfalleh, W., H. Hannachi, N. Tlili, Y. Yahia, N. Nasri, and A. Ferchichi. 2012. Total phenolic contents and antioxidant activities of pomegranate peel, seed, leaf and flower. *Journal of Medicinal Plants Research* 6: 4724-4730.

Elyatem, S.M. and A.A. Kader. 1984. Post-harvest physiology and storage behaviour of pomegranate fruits. *Scientia Horticulturae* 24: 287-298.

Ergun, M. and N. Ergun. 2009. Maintaining quality of minimally processed pomegranate arils by honey treatments. *British Food Journal* 111(4): 396-406.

Esmaillzadeh, A., F. Tahbaz, I. Gaieni, H. Alavi-Majd, and L. Azabakht. 2006. Cholesterol-lowering effect of concentrated pomegranate juice consumption in type II diabetic patients with hyperlipidemia. *International Journal for Vitamin and Nutrition Research* 76: 147-151 [Abstract only].

Ezra, D., I. Kosto, B. Kirshner, M. Hershcovich, and D. Shtienberg. 2014. Heart rot of pomegranate: disease aetiology and the events leading to development of symptoms. *Plant Disease* 99(4): 496-501.

Farber, J.N., L.J. Harris, M.E. Parish, L.R. Beuchat, T.V. Suslow, J.R. Gorney, E.H. Garrett and F.F. Busta. 2003. Microbiological safety of controlled and modified atmosphere packaging of fresh and fresh-cut produce. *Comprehensive Review in Food Science and Food Safety* 2: 142-160.

Fatemi, S., M. Jafarpour, and S. Eghbalsaied. 2012. Study of the effect of Thymus vulgaris and hot water treatment on storage life of orange (Citrus sinensis CV. Valencia). *Journal of Medicinal Plants Research* 6(6): 968-971.

Fawole, O.A. and U.L. Opara. 2012. Composition of trace and major minerals in different parts of pomegranate (*Punica granatum* L) fruit cultivars. *British Food Journal* 114: 1518-1532.

Fawole, O.A. and U.L. Opara. 2013. Effects of storage temperature and duration on physiological responses of pomegranate fruit. *Industrial Crops and Products* 47: 300-309.

Fawole, O.A., N.P. Makunga, and U.L. Opara. 2012b. Antibacterial, antioxidant and tyrosinase-inhibition activities of pomegranate fruit peel methanolic extract. *BMC Complementary and Alternative Medicine* 12: 1-11.

Fawole, O.A., U.L. Opara, and K.I. Theron. 2012a. Chemical and phytochemical properties and antioxidant activities of three pomegranate cultivars grown in South Africa. *Food Bioprocess and Technology* 5: 2934-2940.

Feng, W. and X. Zheng. 2007. Essential oils to control *Alternaria alternata* in vitro and in vivo. *Food Control* 18: 1126-1130 [Abstract only].

Fischer, U.A., R. Carle, and D.R. Kammerer. 2011. Identification and quantification of phenolic compounds from pomegranate (*Punica granantum* L.) peel, mesocarp, aril and differently produced juices by HPLC-DAD-ESI/MSn. *Food Chemistry* 127: 807-821.

Forest, C.P., H. Padma-Nathan, and H.R. Liker. 2007. Efficacy and safety of pomegranate juice on improvement of erectile dysfunction in male patients with mild to moderate erectile dysfunction: a randomized, placebo-controlled, double-blind crossover study. *International Journal of Impotence Research* 19(6): 564-567.

Fuhrman, B., N. Volkova, and M. Aviran. 2005. Pomegranate juice oxidized LDL uptake and cholesterol biosynthesis in macrophages. *Journal of Nutritional Biochemistry* 16: 570-576.

García, E., D.M. Salazar, P. Melgarejo, and A. Coret. 2000. Determination of the respiration index and of the modified atmosphere inside the packaging of minimally processed products. *CIHEAM-Options Mediterraneennes* 247-251.

Ghafir, S.A.M., I.Z. Ibrahim, and S.A. Zaied. 2010. Response of local variety 'Shlefy' pomegranate fruits to packaging and cold storage. *Acta Horticulturae*, 877: 427-432.

Gil, G.I., R. Sanchez, J.G. Marin, and F. Artes. 1996. Quality changes in pomegranate during ripening and cold storage. *Journal European Food Research and Technology* 202: 481-485.

Gil, M.I., F.A. Tomas-Barberan, B. Hess-Pierce, D.M. Holcroft, and A.A. Kader. 2000. Antioxidant activity of pomegranate juice and its relationship with phenolic composition and processing. *Journal of Agricultural and Food Chemistry* 48: 4581-4589.

Glozer, K. and L. Feguson. 2011. *Pomegranate Production in Afghanistan*. UC Regents Davis campus.

Gözlekci, S., S. Ercili, F. Okturen, and S. Sonmez. 2011. Physico-chemical characteristics of three development stages in Pomegranate cv. 'Hicaznar'. Not. Bot. *Notulae Botanicae Horti Agrobotanici Cluj* 39: 241-245.

Graham, S.A., J. Hall, K. Sytsma, and S. Shi. 2005. Phylogenetic analysis of the Lythraceae based on four gene regions and morphology. *International Journal of Plant Sciences* 166: 995-1017.

Groos, K.C., C.Y. Wang, and M. Salveit. 2002. The commercial storage of fruits, vegetables and florist and nursery of crops. *Agricultura Research Service*. Beltsville Area, http://www.ba.ars.usda.gov/hb66/

Haidari, M., M. Ali, S.W. Casscells and M. Madjid. 2009. Pomegranate (*Punica granatum*) purified polyphenol extract inhibits influenza virus and has a synergistic effect with oseltamivir. *Phytomed* DOI:10.1016/j.phymed.2009.06.002 [Abstract only].

Harborne, J.B. 1982. *Introduction to ecological biochemistry*. Academic Press, London.

Hartman, R.E., A. Shah, A.M. Fagan, K.E. Schwetye, M. Parsadanian, R.N. Schulman, M.B. Finn and D.M. Holtzman. 2006. Pomegranate juice decreases amyloid load and improves behavior in a mouse model of Alzheimer's disease. *Neurobiol Disease* 24: 506-515.

Hassan, N.A., A.A. El-Halwagi, and H.A. Sayed. 2012. Phytochemicals, antioxidant and chemical properties of 32 pomegranate accessions growing in Egypt. *World Applied Sciences Journal* 16 (8): 1065-1073.

Hernandez, F., P. Melgarejo, F.A. Tomas-Barberran, and F. Artes. 1999. Evolution of juice anthocyanins during ripening of new selected pomegranate (*Punica granatum*) clones. *European Food Research and Technology* 210: 39-42.

Holland, D., K. Hatip, and I. Bar-Ya'akov. 2009. Pomegranate: Botany, Horticulture and Breeding. In *Horticultural Reviews*, Volume 35, edited by Jules Janick, John Wiley & Sons Inc, 127-191.

Hora, J.J., E.R. Maydew, E.P. Lansky, and C. Dwivedi. 2003. Chemopreventive effects of pomegranate seed oil on skin tumor development in cd1 mice. *Journal of Medicinal Food* 6(3): 157-161.

Howell, A.B. and D.H. Souza. 2013. The pomegranate: effects on bacteria and viruses that influence human health. *Evidence-Based Complement Alternative Medicine* 606212. doi: 10.1155/2013/606212 [Abstract only].

Huang, T.H., G. Peng, B.P. Kota, G.Q. Li, J. Yamahara, B.D. Roufogalis and Y. Li. 2005 Pomegranate flower improves cardiac lipid metabolism in a diabetic rat model: role of lowering circulating lipids. *British Journal of Pharmacology* 145: 767-774.

Husari, A., A. Khayat, H. Bitar, Y. Hashem, A. Rizkallah, G. Zaatari and M.E. Sabban. 2014. Antioxidant activity of pomegranate juice reduces acute lung injury secondary to hyperoxia in an animal model. *BMC Research Notes* 7: 664.

Hussein, Z., O.J. Caleb, K. Jacobs, M. Manley, and U.L. Opara. 2015. Effect of perforation-mediated modified atmosphere packaging and storage duration on physicochemical properties and microbial quality of fresh minimally processed 'Acco' pomegranate arils. *LWT – Food Science and Technology* 64: 911-918.

Ibrahium, M. I. 2010. Efficiency of Pomegranate peel extract as antimicrobial, antioxidant and protective agents. *World Journal of Agricultural Sciences* 6(4): 338-344.

IPGRI. 2001. Regional report CWANA 1999–2000. International Plant Genetic Resources Institute, Rome, Italy, pp. 20-28.

Ishaq, M., M. Usman, M. Asif, and A.I. Khan. 2004. Integrated Pest Management of Mango against Citrus mealybug and Fruit Fly. *International Journal of Agriculture & Biology* 6(3): 452-454.

Ismail, O.M, R. A.A. Younis, and A.M. Ibrahim. 2014. Morphological and molecular evaluation of some Egyptian pomegranate cultivars. *African Journal of Biotechnology* 13(2): 226-237.

ITC, 2015. International Trade Center, http://www.intracen.org/itc/market-insider/fruits-and-vegetables/price-information-updates/ (access on August 07, 2015).

Jafari, A., K. Arzani, E. Fallahi, and M. Barzegar. 2014. Optimizing Fruit Yield, Size, and Quality Attributes in 'Malase Torshe Saveh' Pomegranate through Hand Thinning. *Journal of the American Pomological Society* 68(2): 89-96.

Jafri, M.A., M. Aslam, K. Javed, and S. Singh. 2000. Effect of Punica granatum Linn. (flowers) on blood glucose level in normal and alloxan-induced diabetic rats. *Journal of Ethnopharmacol* 70(3): 309-314 [Abstract only].

James, P.A., S. Oparil, B.L. Carter, W.C. Cushman, C. Dennison-Himmelfarb, J. Handler, D.T. Lackland, M.L. LeFevre, T.D. MacKenzie, O. Ogedegbe, S.C. Smith, L.P. Svetkey, S.J. Taler, R.R. Townsend, J.T. Wright, A.S. Narva and E. Ortiz. 2014. 2014 Evidence-Based Guideline for the Management of High Blood Pressure in Adults. *The Journal of American Medical Association* (5): 507-520.

Jeune, M.A., L.J. Kumi-Diaka, and J. Brown. 2005. Anticancer activities of pomegranate extracts and genistein in human breast cancer cells. *Journal of Medicinal Food*, 8(4): 469-475.

Jung, K.H., M.J. Kim, E. Ha, Y.K. Uhm, H.K. Kim, J.H. Jhung and Y. Sung-Vin. 2006. Suppressive effect of Punica granatum on the production of tumor necrosis factor (Tnf) in BV2 microglial cells. *Biological Pharmacology Bulletin* 29: 1258-1261.

Jurenka, J. 2008. Therapeutic applications of Pomegranate: A review. *Alternative Medicine Review* 13(2): 128-144.

Kader, A.A. 1995. Regulation of fruits physiology by controlled and modified atmosphere. *Acta Horticulturae* 398: 59-70.

Kader, A.A., A. Chordas, and S.M. Elyatem. 1984. Response of pomegranates to ethylene treatment and storage temperature. *California Agriculture Journal* 38: 4-15.

Kahramanoğlu, İ. and S. Usanmaz. 2013. Management strategies for the main fruit damaging pests of pomegranates: Planococcus citri, Ceratitis capitata and Deudorix (Virachola) livia. *African Journal of Agricultural Research* 8(49): 6563-6568.

Kahramanoğlu, İ., S. Usanmaz, and İ. Nizam. 2014. Heart rot infestation on pomegranate fruits and possible transmission vectors in Cyprus. *African Journal of Agricultural Research* 9(10): 905-907.

Karadeniz, F., H.S. Burburlu, N. Koca, and Y. Soyer. 2005. Antioxidant Activity of Selected Fruits and Vegetables Grown in Turkey. *Turkish Journal of Agricultural and Forestry* 29: 297-303.

Kasimsetty, S.G., D. Bialonska, M.K. Reddy, G. Ma, S.I. Khan, and D. Ferreira. 2010. Colon Cancer Chemopreventive Activities of Pomegranate Ellagitannins and Urolithins. *Journal of Agricultural and Food Chemistry* 58(4): 2180-2187.

Kazemi, F., M. Jafararpoor, and A. Golparvar. 2013. Effects of sodium and calcium treatments on the shelf life and quality of pomegranate. *International Journal of Farming and Allied Sciences* 2: 2322-4134.

Kerns, D., G. Wright, and J. Loghry. 2004. Citrus Citrus mealybug (Planococcus citri). The University of Arizona, *Cooperative Extension* 85721, 4p.

Kimball, D. 1991. *Citrus Processing Quality Control and Technology*. Springer Publishing, 470p.

Kohno, H., R. Suzuki, Y. Yasui, M. Hosokawa, K. Miyashita, and T. Tanaka. 2004. Pomegranate Seed Oil Rich in Conjugated Linolenic Acid Suppresses Chemically Induced Colon Carcinogenesis in Rats. *Cancer Science* 95(6): 481-486 [Abstract only].

Köksal, A. I. 1989. Research on the storage of pomegranate (Cv. Gok Bahce) under different conditions. *Acta Horticulturae* 258: 295-302.

Kote, S., S. Kote, and L. Nagesh. 2011. Effect of Pomegranate Juice on Dental Plaque Microorganisms (Streptococci and Lactobacilli). *Ancient Science of Life* 31(2): 49-51.

Ksentini, I., T. Jardak, and N. Zeghal. 2011. First report on *Virachola livia* Klug. (Lepidoptera: Lycaenidae) and its effects on different pomegranate varieties in Tunisia. *EPPO Bulletin* 41(2): 178-182.

Kulkarni, A.P. and S.M. Aradhya. 2005. Chemical changes and antioxidant activity in pomegranate arils during fruit development. *Food Chemistry* 93: 319-324.

Küpper, W., M. Pekmezci, and J. Henze. 1995. Studies on CA-storage of pomegranate fruit (*Punica granatum* L., cv. Hicaz). *Acta Horticulturae* 398: 101-108.

Labbé, M., A. Peria, and C. Saenz. 2010. Antioxidant capacity and phenolic composition of juices from pomegranates stored in refrigeration. In: *International Conference on Food Innovation*, October, 25-29.

Langley, P. (2000). Why a pomegranate? *British Medical Journal*, 321, 1153-1154. http://www.ncbi.nlm.nih.gov/pmc/articles/PMC1118911/ (access on April 30, 2015)

Lansky, E.P., W. Jiang, H. Mo, L. Bravo, P. Froom, W. Yu, N.M. Harris, I. Neeman and M.J. Campbell. 2005. Possible synergistic prostate cancer suppression by anatomically discrete pomegranate fractions. *Invest New Drugs* 23: 11-20 [Abstract only].

Lee, C.J., L.G. Chen, W.L. Liang, and C.C. Wanga. 2010. Anti-inflammatory effects of Punica granatum Linne in vitro and in vivo. *Food Chemistry* 118: 315-322.

Lei, F., X.N. Zhang, W. Wang, D.M. Xing, W.D. Xie, H. Su and L.J. Du. 2007. Evidence of anti-obesity effects of pomegranate leaf extract in high-fat-diet-induced obese mice. *International Journal of Obesity* 31(6): 1023-1029.

Li, Y., C. Guo, J. Yang, J. Wei, J. Xu, and S. Cheng. 2006. Evaluation of antioxidant properties of pomegranate peel extract in comparison with pomegranate pulp extract. *Food Chemistry* 96: 254-260.

Li, Y., S. Wen, B.P. Kota, G. Peng, G.Q. Li, J. Yamahara and B.D. Roufogalis. 2005. *Punica granatum* flower extract, a potent alpha-glucosidase inhibitor, improves postprandial hyperglycemia in Zucker diabetic fatty rats. *Journal of Ethnopharmacol* 99: 239-244 [abstract only].

López-Rubira, V., A. Conesa, A. Allende, and F. Artés. 2005. Shelf life and overall quality of minimally processed pomegranate arils modified atmosphere packaged and treated with UV-C. *Postharvest Biology and Technology* 37: 174-185.

Luangpirom, A., T. Junaimuang, W. Kourchampa, P. Somsapt, and O. Sritragool. 2013. Protective effect of pomegranate (Punica granatum Linn.) juice against hepatotoxicity and testicular toxicity induced by ethanol in mice. *Animal Biology & Animal Husbandry, International Journal of the Bioflux Society* 5(1): 87-93.

Lye, C. 2008. Pomegranate: preliminary assessment of the potential for an Australian industry. *Rural Industries Research and Development Corporation of Australian Government Publication* No. 08/153, P. 17.

Maclean, D., K. Martino, H. Scherm, and D. Hortan. 2011. Pomegranate Production. *University of Georgia Cooperative Extension* Circular 997.

Madrigal-Carballo, S., G. Rodriguez, C.G. Krueger, M. Dreher, J.D. Reed. 2009. Pomegranate (Punica granatum) supplements: authenticity, antioxidant and polyphenol composition. *Journal of Functional Foods* 1: 324-329.

Maguire, K.M., N.H. Banks, and U.L. Opara. 2001. Factors Affecting Weight Loss of Apples. In: *Horticultural Reviews* (edited by Jules Janick) pp. 197-234. John Wiley & Sons, Inc.

Mahajan, P.V., F.A.R. Oliveira, J.C. Montanez, and J. Frias. 2007. Development of user-friendly software for design of modified atmosphere packaging for fresh and fresh-cut produce. *Innovative Food Science and Emerging Technologies* 8: 84-92.

Mahmoudi, E., A. Ahmadi and D. Naderi. 2012. Effect of Zataria multiflora essential oil on Alternaria alternata in vitro and in an assay on tomato fruits. *Journal of Plant Diseases and Protection* 119(2): 53-58.

Malik, A. and H. Mukhtar. 2006. Prostate cancer prevention through pomegranate fruit. *Cell Cycle* 5: 371-373.

Malik, A., F. Afaq, S. Arfaraz, V.M. Adhami, D.N. Syed, and H. Mukhtar. 2005. Pomegranate juice for chemoprevention and chemotherapy of prostate cancer. *Proceedings of the National Academy of Sciences* 102: 14813-14818.

Mamay, M., L. Ünlü, E. Yanık, and A. İkinci, 2014. Infestation and prevalence situation of Carob moth Apomyelois ceratoniae Zell. (Lepidoptera: Pyralidae) in pomegranate orchards in Şanlıurfa. *Turkish Entomology Bulletin* 4(1): 47-54.

Manrakhan, A., C. Kotze, J.H. Daneel, P.R. Stephen, and R.R. Beck. 2013. Investigating a replacement for malathion in bait sprays for fruit fly control in South African citrus orchards. *Crop Protection* 43: 45-53.

Mansour, S.W., S. Sangi, S. Harsha, M.A. Khaleel, and A.R.N. Ibrahim. 2013. Sensibility of male rats fertility against olive oil, Nigella sativa oil and pomegranate extract. *Asian Pacific Journal of Tropical Biomedicine* 3(7): 563-568.

Mansouri, Y., J. Khazaei, and S.R. Hassan-Beygy. 2011. Post-harvest characteristics of pomegranate (Punica granatum L.) fruit. *Agronomical Research in Moldavia* 2: 5-16.

Marpudi, S.L., L.S.D. Abirami, R. Pushkala, and N. Srividya. 2011. Enhancement of storage life and quality maintenance of papaya fruits using Aloe vera based antimicrobial coating. *Indian Journal of Biotechnology* 10: 83-89.

Martinez, J.J., P. Melgarejo, F. Hernandez, D.M. Salazar, and R. Martinez. 2006. Seed characterisation of five new pomegranate (Puica granatum L.) varieties. *Scienta Horticulturae* 110(3): 241-246.

McDonald, A. 2002. Botanical determination of the Middle Eastern tree of life. *Society for Economic Botany* 56: 113-129.

Mehta, R. and E.P. Lansky. 2004. Breast Cancer Chemopreventive Properties of Pomegranate (Punica granatum) Fruit Extracts in a Mouse Mammary Organ Culture. *European Journal of Cancer Prevention* 13(4): 345-355 [Abstract only].

Melgarejo, P., A.C. Sánchez, L. Vázquez-Araújo, F. Hernandez, J.J. Martinez, P. Legua and A.A. Carbonell-Barrachina. 2011. Volatile composition of pomegranates from 9 Spanish cultivars using headspace solid phase microextraction. *Journal of Food Science* 76: 114-120.

Melgarejo, P., D.M. Salazar, and F. Artés. 2000. Organic acids and sugars composition of harvested pomegranate fruits. *European Food Research and Technology* 211: 185-190.

Mena, P., C. García-Viguera, J. Navarro-Rico, D.A. Moreno, J. Bartual, D. Saura and N. Marti. 2011. Phytochemical characterisation for industrial use of pomegranate (*Punica granatum* L.) cultivars grown in Spain. *Journal of the Science of Food and Agriculture* 91: 1893-1906.

Menezes, S.M., L.N. Cordeiro, and G.S. Viana. 2006. *Punica granatum* (pomegranate) extract is active against dental plaque. *Journal of Herbal Pharmacotherapy* 6: 79-92. [Abstract only].

Michailides, T.J., D.P. Morgan, M. Quist, and H.C. Reyes. 2011. Infection of pomegranate by *Alternaria* spp. causing black heart. *University of California publications.*

Miguel, M.G., M.A. Nevesa, and M.D. Antunes. 2010. Pomegranate (*Punica granatum* L.): A medicinal plant with myriad biological properties. *Journal of Medicinal Plants Research*, 4: 2836-2847.

Mirdehghan, S.H. and M. Rahemi. 2005. Effects of hot water treatment on reducing chilling injury of pomegranate (*Punica granatum* L.) fruit during storage. *Acta Horticulturae* 682: 887-892.

Mirdehghan, S.H. and M. Rahemi. 2007. Seasonal changes of mineral nutrients and phenolics in pomegranate (Punica granatum L.) fruit. *Sciencia Horticulturae* 111: 120-127.

Mirdehghan, S.H., M. Rahemi, D. Martinez-Romero, F. Guilléna, J.M. Valverdea, P.J. Zapataa, M. Serranob and D. Valeroa. 2006a. Reduction of pomegranate chilling injury during storage after heat treatment: Role of polyamines. *Postharvest Biology and Technology* 44: 19-25.

Mirdehghan, S.H., M. Rahemi, M. Serrano, F. Guillen, D. Martinez-Romero, and D. Valero. 2006b. Prestorage heat treatment to maintain nutritive and functional properties during postharvest cold storage of pomegranate. *Journal of Agricultural and Food Chemistry* 54: 8495-8500.

Mirjalili, S.A. and E. Poorazizi. 2015. Integrated pest management for carob moth (*Spectrobates ceratoniae* zell.) in Iran. In: *III International Symposium on Pomegranate and Minor Mediterranean Fruits* (eds. Yuan, Z., E. Wilkins, D. Wang, China) [Abstract only]

Morton, J. 1987. Pomegranate. pp. 352–355. In: *Fruits of warm climates*. Miami, FL.

Nabigol, A. and A. Asghari. 2013. Antifungal activity of Aloe vera gel on quality of minimally processed pomegranate arils. *International Journal of Agronomy and Plant Production* 4(4): 833-838.

Nagata, M., M. Hidaka, H. Sekiya, Y. Kawano, K. Yamasaki, M. Okumura and K. Arimori. 2007. Effects of pomegranate juice on human cytochrome P450 2C9 and tolbutamide pharmacokinetics in rats. *Drug Metabolism and Disposition* 35: 302-305.

Nanda, S., D.V.S. Rao, and S. Krishnamurthy. 2001. Effects of shrink film wrapping and storage temperature on the shelf life and quality of pomegranate fruits cv. 'Ganesh'. *Postharvest Biology and Technology* 22: 61-69.

Narayan, T., S. Deshpande, A. Jha, and V.P. Ram Prasad. 2014. Punica granatum (Pomegranate) fruit and its relevance in Oral Hygiene. *IOSR Journal of Dental and Medical Sciences* 13(8): 29-34.

Neifar, M., K. Ellouze-Ghorbel, A. Kamoun, S. Baklouti, A. Mokni, A. Jaouani and S. Ellouze-Chaabouni. 2009. Effective clarification of pomegranate juice using laccase treatment optimized by response surface methodology followed by ultrafiltration. *Journal of Food Process Engineering* 34(4): 1199-1219 [Abstract only].

Nerya, O., A. Gizis, A. Tsyilling, D. Gemarasni, A. Sharabi-Nov, and R. Ben-Arie. 2006. Controlled atmosphere storage of pomegranate. *Acta Horticulturae* (ISHS) 712: 655-660.

Neurath, A.R., N. Strick, Y.Y. Li and A.K. Debnath. 2005. Punica granatum (pomegranate) juice provides an HIV-1 entry inhibitor and candidate topical microbicide. *Annals of the New York Academy of Sciences* 1056: 311-327 [Abstract only].

Newman, R.A. and E.P. Lansky. 2012. *Pomegranate: The Most Medicinal Fruit*. ReadHowYouWant, 168p.

Nizam, İ., İ. Kahramanoğlu and S. Usanmaz. 2015. Effects of different fungicides on the pomegranate heart rot and possible control of fungus infection. *Ciencia E Technica, Vitivinicola* 30(3): 76-89.

Opara, U.L., A.A. Mahdoury, M. Al-Ani, S.A. Al-Khanjari, R. Al-Yahyai, H. Al-Kindi and S.S. Al-Khanjari. 2008. Physiological responses and changes in postharvest quality attributes of 'Helow' pomegranate variety (*Punica granatum* L.) during refrigerated storage. *International Conference of Agricultural Engineering*. (August 31 to September 4).

Özdemir, A.E., E.T. Çandır, M. Kaplankıran, E.M. Soylu, N. Şahinler, and A. Gül. 2010. The effects of ethanol-dissolved propolis on the storage of grapefruit cv. Star Ruby. *Turkish Journal of Agriculture and Forestry* 34: 155-162.

Ozgul-Yucel, S. 2005. Determination of conjugated linolenic acid content of selected oil seeds grown in Turkey. *Journal of the American Oil Chemists' Society* 82: 893-897.

Palou, L., C.H. Crisosto, and D. Garer. 2007. Combination of postharvest antifungal chemical treatment and controlled atmosphere storage to control gray mold and improved storability of Wonderful pomegranates. *Postharvest Biology and Technology* 43: 133-142.

Pantelidis, G., P. Drogoudi, and A. Manganaris. 2012. Physico-chemical and antioxidant properties of pomegranate genotypes in Greece. *Options Mediterraneenes* 335-337.

Pantuck, A.J., J.T. Leppert, N. Zomorodian, W. Aronson, J. Hong, R.J. Barnard, N. Seeram, H. Liker, H. Wang, R. Elashoff, D. Heber, M. Aviram, L. Ignarro and A. Belldegrun. 2006. Phase II study of pomegranate juice for men with rising prostate-specific antigen following surgery or radiation for prostate cancer. *Clinical Cancer Research* 12: 4018-4026.

Park, H.M., E. Moon, A.J. Kim, M.H. Kim, S. Lee, J.B. Lee, Y.K. Park, H.S. Junk, Y.B. Kim and S.Y. Kim. 2010. Extract of Punica granatum inhibits skin photoaging induced by UVB irradiation. *International Journal of Dermatology* 49(3): 276-282 [Abstrtact only].

Parmar, H.S. and A. Kar. 2007. Antidiabetic potential of Citrus sinensis and Punica granatum peel extracts in alloxan-treated male mice. *BioFactors* 31(1): 17-24.

Radunic, M., M.J. Špikaa, S. G. Banb, J. Gadže, J.C. Díaz-Pérez and D. MacLean. 2015. Physical and chemical properties of pomegranate fruit accessions from Croatia. *Food Chemistry* 177: 53-63.

Raga, A. and M.E. Sato. 2005. Effect of spinosad bait against *Ceratitis capitata* (Wied.) and Anastrepha fraterculus (Wied.) (Diptera: Tephritidae) in laboratory. *Neotropical Entomology* 34(5): 815-822.

Ramesh, K.S. and S.F. Shamin. 2012. Role of pomegranate in preventive dentistry. *International Journal of Research in Ayurveda and Pharmacy*. 3(5): 648-649.

Ramezanian, A. and M. Rahemi. 2011. Chilling resistance in pomegranate fruits with spermidine and calcium chloride treatments. *International Journal of Fruit Science* 11: 276-285.

Rashidi, A.A., F. Jafari-Menshadi, A. Zinsaz, and Z. Sadafi. 2013. Effect of Concentrated Pomegranate Juice Consumption on Glucose and Lipid Profile Concentrations in Type 2 Diabetic Patients. *Zahedan Journal of Research in Medical Sciences* 15(6): 40-42.

Reddy, M.K., S.K. Gupta, M.R. Jacob, S.I. Khan and D. Ferreira. 2007. Antioxidant, antimalarial and antimicrobial activities of tannin-rich fractions, ellagitannins and phenolic acids from *Punica granatum* L. *Planta Medicine* 73: 461-467.

Rosenblat, M., N. Volkova, R. Coleman, and M. Aviram. 2006. Pomegranate by Product Administration to Apoiipoprotein e-deficient Mice Attenuates Atherosclerosis Development as a Result on Decreased Macrophage Oxidative Stress and Reduced Cellular Uptake of Oxidized low-density Lipoprotein. *Journal of Agricultural and Food Chemistry* 54(4): 1928-1935.

Rouhani, M., M.A. Samih, H. Izadi, and E. Mohammadi. 2013. Toxicity of new insecticides against pomegranate aphid, Aphis punicae. *International Research Journal of Applied and Basic Sciences* 4(3): 496-501.

Rymon, D. 2012. The prices in Europe of pomegranates and arils. CIHEAM-Options Mediterraneennes. *II International Symposium on the Pomegranate* 103: 37-41.

Saad, H., F. Charrier-El Bouhtoury, A. Pizzi, K. Rode, B. Charrier, and N. Ayed. 2012. Characterization of pomegranate peels tannin extractives. *Industrial Crops Production* 40: 239-246.

Sadeghi, N., B. Jannat, M.R. Oveisi, M. Hajimahmoodi, and M. Photovat. 2009. Antioxidant activity of Iranian pomegranate (Punica granatum L.) seed extracts. *Journal of Agriculture Science and Technology* 11: 633-638.

Sadik, M.S. and M.M.S. Asker. 2014. Antioxidant and antitumor activities of Pomegranate (Punica granatum) peel extracts. *World Journal of Pharmaceutical Sciences* 2(11): 1441-1445.

Salisbury E.J. 1961. *Weeds and Aliens*. London, Collins.

Sarker, M., S.C. Das, S.K. Saha, Z. Al Mahmud, and S.C. Bachar. 2012. Analgesic and Anti-inflammatory Activities of Flower Extracts of *Punica granatum* Linn. (Punicaceae). *Journal of Applied Pharmaceutical Science* 2(04): 133-136.

Saruwatari, A., S. Okamura, Y. Nakajima, Y. Narukawa, T. Takeda, and H. Tamura. 2008. Pomegranate juice inhibits sulfoconjugation in Caco-2 human colon carcinoma cells. *Journal of Medical Foods* 11: 623-628.

Sayed-Ahmed, E.F. 2014. Evaluation of pomegranate peel fortified pan bread on body weight loss. *International Journal of Nutrition and Food Sciences* 3(5): 411-420.

Schubert, S.Y, E.P. Lansky, and I. Neeman. 1999. Antioxidant and eicosanoid enzyme inhibition properties of pomegranate seed oil and fermented juice flavonoids. *Journal of Ethnopharmacology* 66: 11-17.

Seeram, N.P., W.J. Aronson, Y. Zhang, S.M. Henning, A. Moro, R. Lee, M. Sartippour, D.M. Harris, M. Rettig, M.A. Suchard, A.J. Pantuck, A. Belldegrun and D. Heber. 2007. Pomegranate ellagitannin-derived metabolites inhibit prostate cancer growth and localize to the mouse prostate gland. *Journal of Agricultural and Food Chemistry* 55: 7732-7737 [Abstract only].

Sepúlveda, E., L. Galletti, C. Sáenz, and M. Tapia. 2000. Minimal processing of pomegranate var. Wonderful. *CIHEAM-Options Mediterranean* 42: 237-242.

Sethi, V. 1985. Suitability of different packaging material for storing fruit juices and intermediate preserves. *Souvenir of Symposium on Recent Development in Food Packaging*, Mysore, India.

Sezer, E.D., Y.D. Akçay, B. İlanbey, H.K. Yıldırım, and E.Y. Sözmen. 2007. Pomegranate wine has greater protection capacity than red wine on low-density lipoprotein oxidation. *Journal of Medicinal Food* 10(2): 371-374.

Shilikina, L.A. 1973. On the xylem anatomy of the genus *Punica* L. *Botanical Zoology* 58: 1628-1630.

Shwartz, E., I. Glazer, I. Bar-Ya'akov, I. Matityahua, I. Bar-Ilana, D. Hollandb and R. Amir. 2009. Changes in chemical constituents during the maturation and ripening of two commercially important pomegranate accessions. *Food Chemistry* 115: 965-973.

Singh, R.P., A.K. Gupta and A.K. Bhatia. 1990. Utilization of wild pomegranate in North West Himalayas—Status and Problems. *Proceedings of National Seminar on Production and Marketing of Indigenous Fruits*, New Delhi. pp. 100-107.

Singh, S.B. and H.M. Singh. 2000. Bioefficacy and economics of different pesticides against pomegranate butterfly, *Deudorix livia* (Fabricious) (Lycaenidae: Lepidoptera) infesting aonla. *Indian Journal of Plant Protection* 28(2): 173-175.

Spanos, G.A. and R.E. Wrolstad. 1992. Phenolics of apple, pear and white grape juices and their changes with processing and storage—A review. *Journal of Agriculture and Food Chemistry* 40: 1478-1487.

Stover, E. and E.W. Mercure. 2007. The Pomegranate: A New Look at the Fruit of Paradise. *HortScience* 42(5): 1088-1092.

Sturgeon, S.R. and A.G. Ronnenberg. 2010. Pomegranate and Breast Cancer: Possible Mechanisms of Prevention. *Nutrition Reviews* 68(2): 122-128.

Su, T.H. and C.M. Wang. 1988. Life history and control measures for the citrus citrus mealybug and the latania scale insects on grapevine. *Taiwan Plant Protection Bulletin* 30(3): 279-288.

Sumner, M.D., M. Elliott-Eller, G. Weidner, J. Daubenmier, M.H. Chew, R. Marlin, C.J. Raisin and D. Ornish. 2005. Effects of pomegranate juice consumption on myocardial perfusion in patients with coronary heart disease. *American Journal of Cardiology* 96(6): 810-814 [Abstract only].

Syed, D.N., A. Malik, N. Hadi, S. Sarfaraz, F. Afaq, and H. Mukhta. 2006. Photo chemopreventive Effect of Pomegranate Fruit Extract on UVA-mediated Activation of Cellular Pathways in Normal Human Epidermal Keratinocytes. *Photochemistry Photobiology* 82(2): 398-405 [Abstract only].

Tabatabaekoloor, R. and R. Ebrahimpor. 2013. Effect of Storage Conditions on the Postharvest Physico-Mechanical Parameters of Pomegranate (*Punica granatum* L.). *Asian Journal of Science and Technology* 4(5): 82-85.

Tandon, P.L. and B. Lal. 1980. Control of mango citrus mealybug *Drosicha mangiferae* Green by application of insecticides in soil. *Entomology* 5: 67-69.

Tedford, E.C., J.E. Adaskaveg, and A.J. Ott. 2005. Impact of Scholar (a new post-harvest fungicide) on the California pomegranate industry. Online. *Plant Health Progress* doi: 10.1094/PHP-2005-0216-01-PS.

Teixeira da Silva, J.A., T.S. Rana, D. Narzary, N. Verma, D.T. Meshram, and S.A. Ranade. 2013. Pomegranate biology and biotechnology: A review. *Scientia Horticulturae* 160: 85-107.

Toi, M., H. Bando, C. Ramachandran, S.J. Melnick, A. Imai, R.S. Fife, R.E. Carr, T. Oikawa and E.P. Lansk. 2003. Preliminary Studies on the Anti-angiogenic

Potential of Pomegranate Fractions in vitro and in vivo. *Angiogenesis* 6(2): 121-128 [Abstract only].

Türk, G., M. Sönmez, M. Aydın, A. Yüce, S. Gür, M. Yüksel, E.H. Aksu and H. Aksoy. 2008. Effects of pomegranate juice consumption on sperm quality, spermatogenic cell density, antioxidant activity and testosterone level in male rats. *Clinical Nutrition* 27(2): 289-296.

Turkish Statistical Institute, 2015. http://www.turkstat.gov.tr/PreTablo.do?alt_id= 1001 (access on July 07, 2015).

Tzortzakis, N.G., 2007. Maintaining postharvest quality of fresh produce with volatile compounds. *Innovative Food Science and Emerging Technology* 8: 111-116.

Tzulker, R., I. Glazer, I. Bar-Ilan, D. Holland, M. Aviram, and R. Amir. 2007. Antioxidant activity, polyphenol content and related compounds in different fruit juices and homogenates prepared from 29 different pomegranate accessions. *Journal of Agricultural and Food Chemistry* 55: 9559-9570.

UC IPM. 2015. How to manage pests: Pomegranate. University of California, Integrated Pest Management Programme. http://www.ipm.ucdavis.edu/PMG/selectnewpest. pomegranate.html (access on: February 17, 2015)

USAID Report. 2008. Inma Agribusiness Program, Iraq – a Strategy for Pomegranate. http://pdf.usaid.gov/pdf_docs/Pnadp532.pdf (access on February 07, 2014)

USDA. 2015. United States Department of Agriculture, Agricultural Research Service, National Nutrient Database for Standard Reference Release 27. http://ndb.nal.usda.gov/ ndb/foods/show/2486?manu=&fgcd= (access on February 14, 2015)

Valero, M., S. Vegara, N. Martí and D. Saura. 2014. Clarification of Pomegranate Juice at Industrial Scale. *Journal of Food Process and Technology* 5: 324.

Valko, M., D. Leibfritz, J. Moncol, M.T.D. Croninc, M. Mazura and J. Telserd. 2007. Free radicals and antioxidants in normal physiological functions and human disease. *International Journal of Biochemistry and Cell Biology* 39(1): 44-84 [Abstract only].

Varasteh, F., K. Argani, Z. Zamani and A. Mohseni, 2009. Evaluation of the most important characteristics of some commercial pomegranate (*Punica granatum* L.) cultivars grown in Iran. *Acta Horticulturae* 818: 103-108.

Vardin, H. 2000. The possibilities of different pomegranate cultivars grown in Harran plain for using in food industry. Ph.D Diss. Çukurova Univ. Adana, Turkey.

Vardin, H. and H. Fenercioglu. 2009. Study on the development of pomegranate juice processing technology: The pressing of pomegranate fruit. *Proc. Ist Intern. Symposium on Pomegranate*. (Eds A. I. Ozguven) *ISHS Acta Horticulture* 818: 373-381.

Vasconcelos, L.C., M.C. Sampaio, F.C. Sampaio, and J.C. Higino. 2003. Use of *Punica granatum* as an antifungal agent against candidosis associated with denture stomatitis. *Mycoses* 46(5-6): 192-196 [Abstract only].

Vilquez, F., C. Laetreto, and R.D. Cooke. 1981. A study of the production of clarified banana juice using pectinolytic enzymes. *Journal of Food Technology* 16: 115-125.

Viuda-Martos, M., J. Fernández-López, and J.A. Pérez-Álvarez. 2010. Pomegranate and its many functional components as related to human health: A Review. *Comprehensive Reviews in Food Science and Food Safety* 9: 635-654.

Viuda-Martos, M., Y. Ruiz-Navajas, J. Fernandez-Lopez, E. Sendra, E. Sayas-Barbera, and J.A. Perez-Alvarez. 2011. Antioxidant properties of pomegranate (*Punica granatum* L.) bagasses obtained as co-product in the juice extraction. *Food Research International* 44: 1217-1223.

Voravuthikunchai, S.P., T. Sririrak, S. Limsuwan, T. Supawita, T. Iida, and T. Honda. 2005. Inhibitory effects of active compounds from Punica granatum pericarp on verocytotoxin production by enterohemorrhagic Escherichia coli O157: H7. *Journal of Health Science* 51(5): 590-596.

Wang, L. and M. Martin-Green. 2014. Pomegranate and Its Components as Alternative Treatment for Prostate Cancer. *International Journal of Molecular Science* 15: 14949-14966.

Weerakkody, P., J.I. Jobling, M.V. María, and G. Rogers. 2010. The effect of maturity, sunburn and the application of sunscreens on the internal and external qualities of pomegranate fruit grown in Australia. *Scientia Horticulturae* 124: 57-61.

Weller, C. 2014. Medical Daily: Pomegranate May Hold The Key To Quieting Alzheimer's by Chris Weller http://www.medicaldaily.com/pomegranate-compound-could-stem-alzheimers-and-parkinsons-anti-inflammatory-properties-299352 (access on August 26, 2014).

Wetzstein, H., Z. Zhang, N. Ravid, and M.E. Wetzstein. 2011. Characterization of Attributes Related to Fruit Size in Pomegranate. *Hortscience* 46(6): 908-912.

Wetzstein, H.Y. and N. Ravid. 2008. Floral initiation and development in pomegranate. *HortScience* 43: 1140 [Abstract only].

WHO, 2015. Global Health Observatory (GHO) data. http://www.who.int/gho/ncd/risk_ factors/obesity_text/en/ (access on July 29, 2015)

Wilson, E.E. and J.M. Ogawa, 1979. *Fungal, Bacterial, and Certain Nonparasitic Diseases of Fruit and Nut Crops in California*. University of California, Division of Agricultural Sciences, Berkeley, CA, USA, Pub. 4090.

Wren, R.C. 1988. *Potter's new cyclopedia of botanical drugs and preparations*. C.W. Daniel, Essex, UK.

Yasoubi, P., M. Barzegarl, M.A. Sahari, and M.H. Azizi. 2007. Total Phenolic Contents and Antioxidant Activity of Pomegranate (*Punica granatum* L.) Peel Extracts. *Journal of Agricultural Science and Technology* 9: 35-42.

Zahid, N., A. Ali, Y. Siddiqui and M. Maqbool. 2013. Efficacy of ethanolic extract of propolis in maintaining postharvest quality of dragon fruit during storage. *Postharvest Biology and Technology* 79: 69-72.

Zaouay, F., P. Mena, C. Garcia-Viguera and M. Mars. 2012. Antioxidant activity and physico-chemical properties of Tunisian grown pomegranate (*Punica granatum* L.) cultivars. *Industrial Crops and Products* 40: 81-89.

Zarei, M., M. Azizi, and Z. Bashir-Sadr. 2011. Evaluation of physicochemical characteristics of pomegranate (Punica granatum L.) fruit during ripening. *Fruits* 66: 121-129.

Zeweil, H.S., S. El-Nagar, S.M. Zahran, M.H. Ahmed, and Y. El-Gindy. 2013. Pomegranate Peel as a Natural Antioxidant Boosts Bucks' Fertility under Egyptian Summer Conditions. *World Rabbit Science* 21: 33-39.

Zhang, L. and M.J. McCarthy. 2012. Black heart characterization and detection in pomegranate using NMR relaxometry and MR imaging. *Postharvest Biology and Technology* 67: 96-101.

Zhang, Q., Z.M. Radisavljevic, M.B. Siroky, and K.M. Azadzoi. 2010. Dietary antioxidants improve arteriogenic erectile dysfunction. *International Journal of Andrology* 33:1-11.

# INDEX